Bob Weil

Resource Selection by Animals

RESOURCE SELECTION BY ANIMALS

Statistical design and analysis for field studies

Bryan F. J. Manly

Department of Mathematics and Statistics
University of Otago, New Zealand

Lyman L. McDonald

WEST Inc., Wyoming, USA

Dana L. Thomas

Department of Mathematical Sciences
University of Alaska, USA

CHAPMAN & HALL

London · Glasgow · Weinheim · New York · Tokyo · Melbourne · Madras

Published by Chapman & Hall, 2-6 Boundary Row, London SE1 8HN

Chapman & Hall, 2-6 Boundary Row, London SE1 8HN, UK

Blackie Academic & Professional, Wester Cleddens Road, Bishopbriggs, Glasgow G64 2NZ, UK

Chapman & Hall GmbH, Pappelallee 3, 69469 Weinheim, Germany

Chapman & Hall USA, 115 Fifth Avenue, New York, NY 10003, USA

Chapman & Hall Japan, ITP - Japan, Kyowa Building, 3F, 2-2-1 Hirakawacho, Chiyoda-ku, Tokyo 102, Japan

Chapman & Hall Australia, 102 Dodds Street, South Melbourne, Victoria 3205, Australia

Chapman & Hall India, R. Seshadri, 32 Second Main Road, CIT East, Madras 600 035, India

First edition 1993
Reprinted 1995

© 1993 Bryan F.J. Manly, Lyman L. McDonald and Dana L. Thomas

Typeset in 10/12pt Times by Expo Holdings, Malaysia
Printed in Great Britain by the Athenaeum Press Ltd, Gateshead, Tyne & Wear

ISBN 0 412 40140 1

A Catalogue record for this book is available from the British Library

Library of Congress Cataloging-in-Publication Data available

Contents

Preface

We have written this book as a guide to the design and analysis of field studies of resource selection, concentrating primarily on statistical aspects of the comparison of the use and availability of resources of different types. Our intended audience is field ecologists in general and wildlife biologists in particular who are attempting to measure the extent to which real animal populations are selective in their choice of food and habitat. As such, we have made no attempt to address those aspects of theoretical ecology that are concerned with how animals might choose their resources if they acted in an optimal manner.

The book is based on the concept of a resource selection function, where this is a function of characteristics measured on resource units such that its value for a unit is proportional to the probability of that unit being used. We argue that this concept leads to a unified theory for the analysis and interpretation of data on resource selection and can replace many ad hoc statistical methods that have been used in the past.

The book has the following structure: Chapters 1 and 2 provide a review of statistical methods that have been used to study resource selection and some examples of how different types of study can be analysed in terms of resource selection functions; Chapter 3 gives a brief introduction to a range of statistical techniques (mainly used in Chapters 5 to 10) that are useful for modelling data; Chapter 4 deals with the particularly important special case where the resources being studied are in categories of habitat or food types; Chapters 5 and 6 deal with situations where the resource consists of a finite number of identified units and it is known which of these are used and which are not used; Chapters 7 and 8 deal with situations where samples of resource units of different types are taken (e.g. a sample of used units and an independent sample of unused units); Chapter 9 gives an introduction to log-linear modelling in the context of resource selection studies; Chapter 10 gives some suggestions on how to analyse data when the amount of use of resource units is recorded rather than simply whether or not they are used; and, finally, Chapter 11 covers methods for assessing the accuracy of estimates from resource selection functions, as well as differences between these estimates, and ratios of these estimates.

The procedures that we discuss for estimating resource selection functions include some (such as logistic regression) that are standard enough to be available in some of the popular statistical packages, and others (such as fitting a proportional hazards model to sample data) that are by no means standard. Even with standard analyses there are often some complications involved in setting up resource selection data in order to use a computer program to estimate a resource selection function. Therefore, we have found it convenient to produce a computer program that carries out most of the calculations described in Chapters 5 to 8. This program does not cover the methods described in Chapter 4 for categorical data since we feel that these are simple to set up in a spreadsheet program such as LOTUS 123. General log-linear modelling as discussed in Chapter 9 is also omitted since we believe that there are already good computer programs available for this purpose.

The program, which is called RSF (*Resource Selection Functions*), runs on IBM PC and compatible computers with at least 640k of ram. It is written in FORTRAN and uses the MAXLIK algorithm for maximum likelihood estimation of models for count data (Reed and Schull, 1968; Reed, 1969; Manly, 1985, p. 433). For more information contact WEST Inc., 1406 South Greeley Highway, Cheyenne, Wyoming 82007, USA (telephone 307-634-1756; fax 307-637-6981).

We acknowledge the help of our colleagues in writing this book. Particularly, we acknowledge Dan Reed of the Alaska Department of Fish and Game who assisted with early versions of Chapter 4, Ed Arnett, C. R. Bantock, W. F. Harris and Tom Ryder who provided us with raw data from their studies, and Diana Craig who read over the final manuscript. However, any errors or omissions are entirely our responsibility.

BFJM, LLMc and DLT
Denver, January 1992

List of Symbols

Symbol	Definition
I	Number of categories of resource units
m_+	Size of a sample of available resource units
m_i	Number of available units in category i in a sample of available resource units
π_i	Proportion of the population of available units that are in category i
$\hat{\pi}_i$	m_i/m_+, sample proportion of available units in category i
u_+	Size of a sample of used resource units
u_i	Number of units in category i in a sample of used units
o_i	u_i/u_+, sample proportion of used units in category i
f_i	Proportion of the available category i items that are used
$\hat{w}_i$	$o_i/\hat{\pi}_i$, the forage ratio (also called the selection ratio and the preference index)
E_i	$(o_i - \hat{\pi}_i)/(o_i + \hat{\pi}_i)$, Ivlev's electivity index
L_i	$o_i - \hat{\pi}_i$, Strauss' linear selection index
Q_i	$[o_i(1 - \hat{\pi}_i)]/[\hat{\pi}_i(1 - o_i)]$, Jacobs' first selection index
D_i	$(o_i - \hat{\pi}_i)/(o_i + \hat{\pi}_i - 2o_i\pi_i)$, Jacobs' second selection index
α_i	$(o_i/\hat{\pi}_i)/\Sigma(o_i/\hat{\pi}_i)$, Chesson's selection index
B_{i1}	$(u_i/m_i)/\Sigma(u_i/m_i)$, Manly's standardized selection index with used resource units replenished
B_{2i}	$\log(1 - f_i)/\Sigma\log(1 - f_i)$, Manly's standardized selection index with used resource units not replenished
W_i	$f_i/\Sigma f_j$, Vanderploeg and Scavia's first selection index
E^*_i	$(W_i - 1/I)/(W_i + 1/I)$, Vanderploeg and Scavia's second selection index
SI	$MAX[(\Sigma m_i/m_+) - (\Sigma o_i/o_+)]$, Rondorff et al. selection intensity for continuous data
$w^*(x)$	Resource selection probability function for a single period of selection where resource units are characterized by the values $x = (x_1, x_2, ..., x_p)$ for variables X_1 to X_p
$w(x)$	Resource selection function, which is $w^*(x)$ multiplied by an unknown positive constant
$w^*(x, t)$	Resource selection probability function for selection from time 0 to time t
$w(x, t)$	Resource selection function for selection from time 0 to time t

$\phi^*(x, t)$	Probability that a resource unit is not used ('survives') in the time interval 0 to t
$\phi(x, t)$	$\phi^*(x, t)$ multiplied by an unknown positive constant
β_i	Coefficient of the variable X_i in a resource selection or survival function
$\hat{\beta}_i$	Estimate of β_i
$se(\hat{\beta}_i)$	Standard error of the estimator of β_i
X_L^2	Log-likelihood chi-squared statistic for measuring the goodness of fit of a model for data
X_P^2	Pearson chi-squared statistic for measuring the goodness of fit of a model for count data
R_i	Standardized residual for the ith observation in a set of data
μ_x	Population mean of the random variable X
s_X	Sample standard deviation of the variable X
r_{XY}	Sample Pearson correlation between variables X and Y
$var(X)$	Variance of the random variable X
$cov(X, Y)$	Covariance of random variables X and Y
A_+	Size of a finite population of available resource units
A_i	Number of the A_+ units that are in category i
U_i	Number of used resource units in category i for the population
U_+	Number of used resource units in all categories
w_i^*	Proportion of the A_i available resource units that are used
$\hat{w}_i^*$	Estimate of w_i^*
z_α	Value that is exceeded with probability α by a standard normal random variable
$E(X)$	Expected (mean) value of a random variable X
u_{ij}	Number of category i resource units used by animal j
u_{i+}	Number of category i resource units used by all sampled animals
u_{+j}	Number of resource units used by animal j
u_{++}	Total number of units used by all sampled animals
$\hat{w}_{ij}$	Selection ratio for the jth animal and the ith resource category
π_{ij}	Proportion of the resources available to animal j that are in category i
$\hat{\pi}_{ij}$	Sample estimate of π_{ij}
S	Number of selection episodes
Θ_{ij}	Probability that a type i resource unit will be used in the time interval (t_{j-1}, t_j)
P_a	Probability of sampling an available resource unit
$P_u(t)$	Probability of sampling a used resource unit at time t
$P_{\bar{u}}(t)$	Probability of sampling an unused resource unit at time t
$u_i(t)$	Number of resource units of type i found in a sample of used units taken at time t
$\bar{u}_i(t)$	Number of resource units of type i found in a sample of unused units taken at time t

P_u	Probability of sampling a used resource unit when only one sample of these units is taken
$P_{\bar{u}}$	Probability of sampling an unused resource unit when only one sample of these units is taken
μ	Population mean vector
Σ	Population covariance matrix
$\Omega(x)$	A function such that selecting individuals with measurements $x = (x_1, x_2, ..., x_p)$ with probabilities $\Omega(x)$ from one multivariate population will produce a second population of a specified type
ρ	Proportion of available units that are used

1 Introduction to resource selection studies

In this chapter we provide motivation for the study of resource selection, define terms, discuss study designs and sampling, and give an historical perspective on the statistical evaluation of resource selection.

1.1 MOTIVATION AND DEFINITIONS

Adequate quantities of usable resources are necessary to sustain animal populations. Therefore, biologists often identify resources used by animals and document the availability of those resources. The need for such documentation is especially critical in efforts to preserve endangered species and manage exploited populations. Determining which resources are selected more often than others is of particular interest because it provides fundamental information about the nature of animals and how they meet their requirements for survival.

Differential resource selection is one of the principal relationships which permit species to coexist (Rosenzweig, 1981). It is often assumed that a species will select resources that are best able to satisfy its life requirements, and that high quality resources will be selected more than low quality ones. The availability of various resources is not generally uniform, and use may change as availability changes. Therefore, used resources should be compared to available (or unused) resources in order to reach valid conclusions concerning resource selection.

When resources are used disproportionately to their availability, use is said to be selective. We define the usage of a resource as that quantity of the resource utilised by an animal (or population of animals) in a fixed period of time. The availability of a resource is the quantity accessible to the animal (or population of animals) during that same period of time. We distinguish between availability and abundance by defining the latter as the quantity of the resource in the environment. Although selection and preference are often used as synonyms in the literature, we define them differently. To us, selection is the process in which an animal chooses a resource, and preference is the likelihood that a resource will be selected if offered on an equal basis with others (Johnson, 1980).

Resource selection occurs in a hierarchical fashion from the geographic range of a species, to individual home ranges within a geographical range, to use of general features (habitats) within the home range, to the selection of particular elements (food items) within the general features (or feeding site). The criteria for selection may be different at each level (Johnson, 1980; Wiens, 1981) so that, when making inferences, researchers studying selection must keep in mind the level being studied.

Most commonly, selection studies deal with food or habitat selection. Food selection may be among various prey species or among sizes, colours, shapes, etc. of the same species. Habitat selection may be among various discrete habitat categories (e.g., open field, forest, rock outcropping) or among a continuous array of habitat attributes such as shrub density, percentage cover, distance to water, canopy height, etc. Thus, the variables observed in a selection study may be discrete, or continuous or some combination of the two.

Considerable variation exists in the motivation for conducting resource selection studies. For example, there is sometimes a need to provide quantitative information that is indicative of the long-term resource requirements of a population, as for the debate concerning whether old growth forest is vital to the continued existence of the spotted owl (*Strix occidentalis*) in the Pacific Northwest of the United States (Laymon *et al.*, 1985; Forsman *et al.*, 1984) or for the existence of black-tailed deer (*Odocoileus hemionus*) on Admiralty Island, Alaska (Schoen and Kirchhoff, 1985). Resource selection studies are commonly carried out for this reason. However, it should be noted that a resource item may be highly favoured but if it is difficult to find then it cannot be utilized much. Conversely, if resources which are less favoured are the only ones available then they may of necessity comprise a large proportion of those used (Petrides, 1975, Chapter 8; White and Garrott, 1990). Therefore, researchers should proceed cautiously when using the results of selection studies to determine the relative importance of resources.

Resource selection studies are often used in fisheries. Gear selectivity has to be allowed for in estimating abundance, describing population size structures and evaluating mortality rates (Hamley and Regier, 1973; Beamsderfer and Rieman, 1988; Millar and Walsh, 1992; Walsh *et al.*, 1992). These studies fall under the present theory because sampling gear is sometimes 'biased' in that the items available in the population are not selected in proportion to their availability. In addition, researchers often investigate river characteristics such as the water velocity or depth that are selected by fish. Examples are Parsons and Hubert's (1988) study of spawning site selection by kokanees (*Oncorhynchus nerka*) and the study of Belaud *et al.* (1989) of brown trout (*Salmo trutta*) stream site selection at various stages of development. The latter studies provide information from in-stream flow incremental methods (Bovee, 1981) to evaluate habitat use.

Another use of resource selection studies is as part of modelling and projecting the impact of habitat change. Under certain assumptions the ratio of animal densities equals the ratio of resource availabilities for any two habitats at equilibrium (Fagen, 1988). These relationships are used to calculate relative habitat values from habitat selection studies and to define hypothetical carrying capacity curves based on habitat values under various conditions. Such an approach was used by Schoen and Kirchhoff (1985) to model deer production after logging. This approach is quite controversial, as evidenced by the differences of opinion stated by Fagen (1988), van Horne (1983) and Hobbs and Hanley (1990). Other studies utilizing selectivity to evaluate the effects of human disturbance include the study of Edge *et al.* (1987) of elk (*Cervus elaphus*) habitat selection and Edwards and Collopy's (1988) study of osprey (*Pandion haliaetus*) nest trees. In such studies undisturbed sites often serve to provide baseline information which helps managers to evaluate the impact of man on animals.

Other situations where resource selection studies have a major role include the evaluation of the effect of domestic animals on wild animal forage, such as Bowyer and Bleich's (1984) study of the effects of cattle grazing on selected habitats of southern mule deer (*Odocoileus hemionus*); the evaluation of the Habitat Suitability Indices (HSI) that are used by the United States Fish and Wildlife Service (1981) to characterize habitat quality for selected species, such as the test devised by Thomasma *et al.* (1991) of the HSI for the fisher (*Martes pennanti*); estimating parameters in optimal foraging and diet prediction models (Pyke *et al.*, 1977; Nelson, 1978); and the determination of the specific prey cues that result in predation, such as Zaret and Kerfoot's (1975) study of fish predation on *Bosmina longirostris* to determine whether selection was related to body-size or visibility. We also note that early work on habitat selection was closely associated with ideas on speciation, niche theory, and range expansion. Although habitat selection in itself is no longer generally considered a major factor in speciation for birds and mammals, some entomologists have suggested that food and oviposition selection play a significant role in evolution and speciation among some insects (Feder *et al.*, 1990; Thompson, 1988).

Many factors, including population density, competition with other species, natural selection, the chemical composition or texture of forage, heredity, and predation contribute to resource selection (Peek, 1986) and numerous models and theories of resource selection have been proposed that incorporate subsets of these factors. These include foraging models (Emlen, 1966; Rapport and Turner, 1970; Werner and Hall, 1974; Ellis *et al.*, 1976; Pyke *et al.*, 1977; Rapport, 1980; Nudds, 1980, 1982; Belovsky *et al.*, 1989) and habitat selection models (Bryant, 1973; Rosenzweig, 1981; Whitham, 1980). We do not delve into such models in this book. Instead, we restrict attention to statistical techniques for the detection and measurement of the degree to which a resource is selected or avoided.

The reasons why a particular resource is selected or avoided is not directly revealed by the estimation of the amount of use or avoidance. We cannot be certain, for example, that any food item is distasteful just because it is rarely used. It may be the case, but there is no way of knowing from availability and use data alone that this is true. However, if we learn that there is selection for or against a resource then this is a starting point for further in-depth study (Petrides, 1975).

1.2 THE DATA FOR RESOURCE SELECTION STUDIES

Throughout this book we will make the assumption that the resource being studied can be considered to consist of a number of discrete resource units and that data are collected from a census or sampling of these units. The complete set of these units will be referred to as the universe of available resource units. This division of the resource into units will occur naturally in some cases, such as when the resource units are individual prey items. At other times the division must be imposed by the researcher. This would be the case, for example, if a study area is divided into quadrats that may or may not be used by the individuals in an animal population. Such a division of the study area into grid cells has been recommended by Porter and Church (1987).

Studies sometimes include the identification of individual animals. For example, Gionfriddo and Krausman (1986) used individually identified radio-collared sheep (*Ovis canadensis mexicana*) to study their summer habitat use. However, individual moose (*Alces alces*) were not identified by Neu *et al.* (1974) when they examined the use of areas in each of four burn categories during several aerial surveys. Similarly, studies on food habits may use collected individuals, as in Hohman's (1985) study of ring-necked ducks (*Aythya collaris*), or methods that do not distinguish individuals, such as the use, by Keating *et al.* (1985), of bighorn sheep (*Ovis canadensis canadensis*) pellets to study winter food habits.

Studies may involve classifying observations into resource categories, or measures of specific variables characteristic of those resources may be obtained. For example, Murphy *et al.* (1985) collected habitat use data for white-tailed deer (*Odoncoileus virginianus*) by classifying radio locations into one of six habitat types. The proportion of radio locations in each habitat type was then compared to the relative availability of the respective habitat type in the study area. On the other hand, Dunn and Braun (1986) examined the selection of habitat by juvenile sage grouse (*Centrocerus urophasianus*) by comparing habitat attribute data such as the shrub density and the distance to another cover type for radio locations and random sites.

A wide variety of methods can be used to collect data. For example, resource availability may be evaluated accurately from a map (Neu *et al.*, 1974) or sampled by randomly selecting sites (Marcum and Loftsgaarden, 1980). In summary, then, it can be said that the data used to evaluate resource selection may be collected through a census or by sampling with one or more procedures, may be categorical or continuous, and may be univariate or multivariate.

1.3 SAMPLING DESIGNS

Given the decision to study resource selection for a particular animal species, the next decision is to determine the scale or scales of selection to focus on (Johnson, 1980). In making this decision the researcher must use what is known about the biology of the animal. For example, if the animal being studied is territorial then selection is commonly studied on a different scale from what would be used for a non-territorial animal (Johnson, 1980). Also, in studying nest site selection it is important to know whether animals always return to the same area. If they do then this has an impact on what resources can be assumed to be available at the scale of interest.

In some cases determining the scale of selection will be a major goal of a study, and as a general rule researchers should consider studying selection at more than one scale. For example, Danell *et al.* (1991) investigated whether moose select forage at the individual tree level or on patches of trees by providing moose with access to artificial stands of trees for which the mixture of forage species and their spacings were controlled. Field studies that cannot manipulate resources in this way may still examine scale questions by measuring availabilities at various distances from used sites (Larsen and Bock, 1986).

Another initial decision in designing a resource selection study is the choice of the study area and its boundaries. This choice may have a significant impact on the results of subsequent data analysis, especially if resource units are arranged in an aggregated pattern (Porter and Church, 1987). When choosing a study area the researcher must consider the distribution of resource units, the scale of selection studied, what is truly available to the animals, and manpower and budget constraints for sampling.

Three general study designs for evaluating selection have been identified in the literature (Thomas and Taylor, 1990). These study designs differ with respect to the level at which resource use and availability are measured: at the population level or for each animal. They are defined as follows:

1.3.1 Design I

With this design, measurements are made at the population level. Used, unused, or available resource units are sampled or censused for the entire study area and for the collection of all animals in the study area. Individual animals are not identified. Examples of this design are:

(1) Aerial or ground surveys such as line transects are used to classify animal locations into resource categories (habitat or forage types). Either maps or aerial photography are used to census availability or random plots are sampled to estimate availability. The percentage use for each category is then compared to its respective availability to evaluate selection. For example, Stinnett and Klebenow (1986) examined cover-type selection of California quail (*Callipepla californica*) by classifying flushes observed during ground surveys into cover types. Maps and aerial photography were partitioned into the respective cover types to evaluate availability. This approach has also been used to study diet selection by comparing the percentage of several vegetation types that were browsed to the percentage

available on plots or transects randomly located in the study area (Lagory *et al.*, 1985).

(2) Randomly located plots are classified as used or unused on the basis of signs such as pellets or tracks. Plot attributes such as the percentage of vegetation cover, the shrub density, the distance to water etc. are measured for each sampled plot, and used and unused sites are compared to evaluate selection. In fisheries studies, nets are set in a sample of locations and the number of fish in each location measures the use of that location. Locations are classified into categories such as shallow or deep, with a fast or slow current, or attributes are measured at each sampled location. Available or unused locations are then compared with used locations.

1.3.2 Design II

With this design, individual animals are identified and the use of resources is measured for each, but availability is measured at the population level. Some examples are:

(1) A sample of animals is collected or otherwise identified via neck collars, radioactive tracing, ear tags, radio transmitters, or coloured leg bands, and the resource units used by each animal are recorded. Each used resource unit may be classified into a category or some attributes are measured on it. The collection of available resource units is sampled or censused (e.g., using aerial photography, geographic information systems, or maps) for the entire study area.

(2) Individual animals are identified and their home range determined. The proportions of resource types in home ranges (as determined by sampling or the partitioning of maps) are compared to the proportions in the entire study area. For example, Roy and Dorrance (1985) compared habitat availabilities within coyote (*Canis latrans*) home ranges with the availabilities in the entire study area. This approach has been criticized by White and Garrott (1990, p. 201) on the grounds that the home range represents a prior selection of habitat.

(3) The relative number of relocations of a marked individual in each habitat type is compared to the proportion of that habitat in the study area. These data are then replicated for each marked animal. This method has been used for example in studies addressing habitat selection by sheep (Gionfriddo and Krausman, 1986) and elk (McCorquodale *et al.*, 1986).

(4) Foraging ecology studies often compare stomach or faecal contents of identified individuals with random samples of available food from the entire study area (Prevett *et al.*, 1985).

When only a single observation of use is made for each animal there is no way of distinguishing between design I and design II. For example, this is the case in studies of nest or bed site selection (Petersen, 1990; Huegel *et al.*, 1986) when only one site is identified for each animal. Thus, design II studies with only one observation of use per animal are a special case of design I studies.

1.3.3 Design III

With this design, individuals are identified or collected as in design II, and at least two of the sets (used resource units, unused resource units, available resource units) are sampled or censused for each animal. Some examples are:

(1) The animals in a sample are radio-collared, and the relocations of an animal identify used resource units for that animal. The used units are either classified into types or attributes are measured on each. The collection of available or unused resource units within each animal's home range is sampled or censused.

(2) A sample of animals is observed feeding for some fixed time period. Individuals are then collected and stomach analysis performed on each. Prey items are sorted by type (e.g., species or colour), or measurements (length, volume) are taken on each prey item. A sample of available (or unused) prey items is taken at the feeding site of each animal collected.

(3) The home range or territory of an individual animal is determined, and the use and availability of resources are compared within that area. For example, Rolley and Warde (1985) used radio-telemetry to identify home ranges and the use of various vegetation types by individual bobcats (*Felis rufus*). LANDSAT data were then used to estimate the relative proportions of each vegetation type within each home range. Foraging studies using this design typically collect individuals for gut analyses and measure food availability at each collection site (Hohman, 1985).

Designs II and III each involve uniquely identified individuals. Therefore, making inferences for the population of animals requires that we assume that the animals identified are a random sample from that population. The sample design then becomes a several stage process: selection of a sample of animals, selection of samples of used and available resource units for each animal, and (often) subsampling of the chosen resource units.

Designs II and III allow an analysis of resource selection for each individual animal. Hence estimates calculated from observations on individual animals may be used to estimate parameters for the population of animals and estimates of variability of these estimates. This approach, which is called first and second stage analysis by Cox and Hinkley (1974), has been recommended for analysis of resource selection data by White and Garrott (1990) and was used by Porter and Labisky (1986) in their study of foraging habitat selection of red-cockaded woodpeckers (*Picoides borealis*) with pairs and groups of birds as first stage units.

The advantages of this type of approach are:

(a) The observations on any one animal may be time-dependent. For example, the independence of several radio relocations depends on the time between these relocations and the animal's diurnal behaviour pattern. Similarly, in food studies the selection of consecutive prey items may be dependent. Therefore, as a general rule it is better to estimate sampling variances and test hypotheses using variation between animals rather than the variation between observations on one animal. In effect this means that inferences become design-based (relying on random sampling of animals) rather than model-based (relying on the assumed statistical model being correct). The advantage of design-based inference over model-based inference is that the former is far less dependent on the assumed statistical model being correct.

(b) It allows estimation techniques that are applicable at the population level for design I to be applied to individuals with designs II and III.

(c) The variation among individuals may be examined to determine, for example, whether the selection for one resource is more variable than for others, whether sex or age differences between animals occur, and to identify animals that are unusual with respect to their selectivity.

These advantages may also be obtained by treating groups of animals as first-stage units providing that the groups used in the study are a random sample from the population of groups.

In addition to the field study designs described above, controlled experimental studies have also been conducted to evaluate resource selectivity. This approach for the evaluation of forage selection is fairly common, with examples being the studies of Colgon and Smith (1985) and Hohf *et al.* (1987). Habitat manipulation studies are less common but can be implemented (White and Trudell, 1980; Munro and Rounds, 1985). Our emphasis in this book is on observational studies; however, we do provide several examples of experimental studies and their analysis.

1.4 SAMPLING PROTOCOLS

Some resource units such as food items can be used only once, while others, such as habitat units, may be used repeatedly. In either case there may be resource units that are unused by a particular definition over the period of a study. We can therefore partition the population of resource units into two sets consisting of used units and unused units. Resource selection may be detected and measured by comparing any two of the three possible sets (used, unused, and available) of resource units. On this basis, three sampling protocols can be identified depending on the two sets measured.

With sampling protocol A (SPA), available resource units are either randomly sampled or censused, and a random sample of used resource units is taken. With sampling protocol B (SPB), available resource units are randomly sampled or censused, and a random sample of unused resource units is taken. With sampling protocol C (SPC) unused resource units and used resource units are both independently sampled. Each of these sampling protocols may be implemented for each of the sampling designs I, II and III described in the previous section, and the particular combination of design and protocol used to gather data determines some of the underlying assumptions required for a subsequent analysis (e.g., whether the availability of each resource is the same for all animals).

In addition to the three sampling protocols there are cases where a complete census of used and unused resource units can be made. The evidence for selection can then come from considering the variation implied by a stochastic model for the selection process rather than from the consideration of random sampling variation.

In some designs the samples of used and available resource units are not independent. Thus in Hohman's (1985) study of ring-necked ducks, samples of available food items were taken specifically at the feeding site where birds were collected after they had been observed to feed for a short period. The use and availability samples are then paired by location. Similarly, in the study of Huegel *et al.* (1986) of bedsite selection by white-tailed deer fawns, habitat characteristics were measured at the bedsite and at an adjacent plot. Again, the measurements are paired by location.

Clearly, the analysis of data should take into account whether the samples taken of different types of resource unit (used, unused and available) are independent or paired. Many statistical techniques commonly employed to evaluate resource selection require the assumption that samples being compared are independent. If in fact the samples are paired then this assumption is violated and the results of such an analysis are suspect.

1.5 INDICES OF SELECTION

Early researchers simply described their findings on resource use and availability. Initially, quantification of resource selection occurred in the analysis of food studies. For example, see Kalmback's (1934) review of stomach analysis publications. Most of these early studies only indicated the number of animals consuming each prey item and the percentage consumption. Variability among animals and locations made it difficult for researchers to compare their results because differences were assessed subjectively.

Scott (1920) is commonly cited (e.g., Cock, 1978; Pearre, 1982) as the first author to quantify selection. He divided the average number of each prey species per fish stomach per unit of time by the number found in plankton hauls per unit

Table 1.1 Commonly cited indices of selectivity[*]

References	Index
Savage (1931)	$w_i = o_i/\hat{\pi}_i$, the forage ratio
Ivlev (1961)	$E_i = (o_i - \hat{\pi}_i)/(o_i + \hat{\pi}_i)$, Ivlev's electivity index
Strauss (1979), Jolicoeur and Brunel (1966), Ready *et al.* (1985)	$L_i = o_i - \hat{\pi}_i$, Strauss' linear index
Jacobs (1974)	$Q_i = [o_i(1 - \hat{\pi}_i)]/[\hat{\pi}_i(1 - o_i)]$ and $D_i = (o_i - \hat{\pi}_i)/(o_i + \hat{\pi}_i - 2o_i\hat{\pi}_i)$, also $\log(Q_i)$ and $\log(D_i)$
Chesson (1978), and Paloheimo (1979)	$\alpha_i = (o_i/\hat{\pi}_i)/\Sigma(o_i/\hat{\pi}_i)$, Chesson's index
Manly *et al.* (1972)	$B_{i1} = (u_i/m_i)/\Sigma(u_i/m_i)$ and
Manly (1973, 1974)	$B_{i2} = \log(1 - f_i)/\Sigma\log(1 - f_i)$
Vanderploeg and Scavia (1979a,b)	$W_i = f_i/\Sigma f_i$ and $E^{*}_i = (W_i - 1/I)/(W_i + 1/I)$
Bowyer and Bleich (1984)	Importance $= o_i\hat{\pi}_i$
Rondorff *et al.* (1990)	$SI = MAX [(\Sigma m_i/m_+) - (\Sigma o_i/o_+)]$, selection intensity for continuous data

[*]m_+ = size of the sample of available resource units; m_i = number of available units in category i in the sample (i = 1, 2, ..., I); $\hat{\pi}_i = m_i/m_+$ = sample proportion of available units in category i; u_+ = size of the sample of used resource units; u_i = number of units in category i in the sample of used units; $o_i = u_i/u_+$ = sample proportion of used resource units in category i; and f_i = proportion of the initial number of category i items that are used.

For B_{i1}, m_i is the maintained number of resource i items available, kept constant by replacing resource units as they are used (if necessary). If items are replaced as used then this index is equivalent to Chesson's index. If items are not replaced then this index is equivalent to Vanderploeg and Scandia's index. The index B_{i2} is intended for field data without resource replacement.

area. Thus, this first index used the ratio of the rate of consumption of a prey type to the density at which it was present (Cock, 1978).

The commonly employed ratio of percentage use divided by percentage available appears to have been independently proposed by Savage (1931) (Cock, 1978), three Russian researchers including A. A. Shorygin (Ivlev, 1961), and by Hess and Swartz (1940) who referred to it as the forage ratio. Ivlev (1961) found the 0 to infinity range of the forage ratio to be cumbersome and proposed an alternative electivity index with the range −1 to +1. Numerous other indices have also been proposed, as shown in Table 1.1.

Several other authors have suggested indices related to selectivity, including Pearre (1982), Holmes and Robinson (1981), Rachlin *et al.* (1987), Pinkas *et al.* (1971), and Owen-Smith and Cooper (1987). Pearre (1982) gave two selectivity indices based on chi-squared statistics that compare one resource at a time to all other resources combined. Holmes and Robinson (1981) used a 'tree preference index' to study tree species preferences of foraging insectivorous birds by summing percentage deviations of bird foraging frequencies in various tree species from relative importance values of the trees. The relative importance values were calculated from the densities, frequencies of occurrence and basal areas per hectare for individuals of each tree species greater than 2.5 cm in

Table 1.2   Lechowicz's (1982) seven criteria for his optimal index $f(o_i, \hat{\pi}_i)$

1. Random model	$f(o_i, \hat{\pi}_i) = 0$ if, and only if, $o_i = \hat{\pi}_i$
2. Symmetry	If $o_i = \hat{\pi}_i$ then $|f(o_i + c, \hat{\pi}_i)| = |f(o_i - c, \hat{\pi}_i)|$, where c is any constant
3. Range	For any number of resources, MAX $f(o_i, \hat{\pi}_i) = f(1.0, \hat{\pi}_i)$ and MIN $f(o_i, \hat{\pi}_i) = f(0, \hat{\pi}_i)$, i.e., the indices maximum should occur when only one resource is used, and the minimum should occur when a resource is not used
4. Linearity	$f(o_i, \hat{\pi}_i) - f(o_i + a, \hat{\pi}_i) = b$ for any o_i and $\hat{\pi}_i$, where a and b are constants
5. Robustness	An index should not be sensitive to sampling errors, particularly for rare or little-used resources
6. Testability	An index should be amenable to statistical comparisons between subgroups (e.g., sexes, age groups, etc.) or between samples (times, locations)
7. Stability	An index should give comparable results for samples from sites differing in type or abundance of available resources

diameter at breast height. In assessing the food selection of ungulates, Owen-Smith and Cooper (1987) proposed two indices of acceptance: site-based acceptance, calculated as the number of 30-minute intervals during which the resource was used, divided by the number of 30-minute intervals during which the resource was present within 10 m; and plant-based acceptance, calculated as the number of individual plants of type i eaten, divided by the number of plants of that type encountered within neck reach.

Indices of selection have been reviewed by Krueger (1972), Cock (1978), Strauss (1979), Loehle and Rittenhouse (1982), Pearre (1982), and Lechowicz (1982). Lechowicz evaluated indices according to the seven criteria that are defined in Table 1.2, where these pertain to his idea of an optimal index. Pearre classified the indices that he reviewed into two types: those reflecting selection for the particular circumstance observed ($\hat{w}$, E, D, Q, L) and those measuring an invariant degree of preference (W, E^*, α, B_1, B_2). The former group (with the exception of $\hat{w}$) is a collection of ad hoc methods that do not estimate any biologically meaningful value, while the latter collection attempt to estimate the probability (or some multiple of the probability) that the next resource used will be of a specific type. Because of their biological interpretation we prefer the latter indices, which relate directly to the concept of a resource selection function as discussed in later chapters of this book.

1.6 HYPOTHESIS TESTS AND CONFIDENCE INTERVALS

Hypothesis tests provide a structured means of making decisions about a population using sample data for which probabilities of errors can be evaluated. In resource selection studies tests can be used to determine objectively whether resources are being used selectively and to compare the strength of selectivity

Table 1.3 Hypothesis tests used to evaluate resource selection

Statistical test	Example references
Categorical data	
Chi-square goodness-of-fit test	Neu et al. (1974), Byers et al. (1984)
Johnson's prefer method	Johnson (1980)
Friedman's test	Pietz and Tester (1982, 1983)
Chi-squared test of homogeneity	Marcum and Loftsgaarden (1980)
Quade's test	Alldredge and Ratti (1986, 1992)
Log-linear models	Heisey (1985)
Wilcoxon's signed rank test	Kohler and Ney (1982), Talent et al. (1982)
Continuous data	
Kolmogorov–Smirnov two-sample test	Raley and Anderson (1990), Petersen (1990)
Multiple regression	Lagory et al. (1985), Grover and Thompson (1986), Giroux and Bedard (1988), Porter and Church (1987)
Logistic regression	Thomasma et al. (1991), Hudgins et al. (1985)
Discriminant function analysis	Dunn and Braun (1986), Edge et al. (1987), Rich (1986), Dubuc et al. (1990)
Multivariate analysis of variance	Stauffer and Peterson (1985)
Principal components	Edwards and Collopy (1988)
Geometric method	Kincaid and Bryant (1983)
Multiple response permutation procedures (e.g., Mielke, 1986)	Alldredge et al. (1991)

among resources. Numerous statistical tests for evaluating overall resource selectivity have been employed, as shown in Table 1.3, and, in addition, several confidence interval procedures and tests that consider one resource at a time have been used to assess selectivity. These latter tests include procedures described by Hobbs and Bowden (1982), Talent et al. (1982), and Iverson et al. (1985), as well as intervals for many of the selectivity indices listed in Table 1.1.

The tests and confidence intervals for single-resource types provide a means of assessing the variability due to sampling error but commonly do not make use of the natural multivariate nature of selectivity data. Furthermore, as noted by Thomas and Taylor (1990), the overall type I error rate for all resource types is not controlled. Thomas and Taylor also point out that the choice of study design often restricts the type of analysis that can be conducted, and that some misuses have occurred in the past.

Comparisons of several of the tests that are listed in Table 1.3 have been carried out by Alldredge and Ratti (1986, 1992). They used simulations of data from design II studies, with availability censused and use sampled, to compare the chi-squared goodness-of-fit test, Johnson's (1980) prefer method, the Friedman (1937) test, and Quade's (1979) test with respect to the null hypotheses tested, the assumptions required, and type I and II error rates. They concluded that no single method was superior for both types of error.

1.7 DISCUSSION

The above review indicates that a unified statistical theory is needed for the analysis of resource selection studies. This is provided by the theory presented in this book, which is based on the concept of a resource selection function, which is a function such that its value for a resource unit is proportional to the probability of that unit being used. The resource selection function will usually be dependent on several characteristics measured on the resource units. If it is specialized to a single categorical variable then the indices of Manly *et al.* (1972), Manly (1973, 1974) and Chesson (1978) are obtained. This case is considered in Chapter 4. Selection functions of multiple characteristics are more general and are considered in Chapters 5 to 11. Statistical inferences are made by design-based methods from the random sampling variation, or by model-based methods where evidence for selection comes from variation about a stochastic model.

2 Examples of the use of resource selection functions

This chapter provides an overview of the use of resource selection probability functions for studying resource selection. Examples are presented to illustrate (a) the commonly occurring case where resources are considered to be in several categories, (b) the difference between census and sample data, and (c) the difference between studies that involve one period of selection time and studies that involve several periods of selection time. In addition, the assumptions that are required for the estimation of resource selection functions are discussed.

2.1 INTRODUCTION

A general approach to the study of resource selection can be based on the concept of a resource selection probability function, where this is a function which gives probabilities of use for resource units of different types. This approach can be used whenever the resource being considered can be thought of as consisting of a universe of N available units, some of which are used and the remainder unused, and where every unit can be characterized by the values that it possesses for certain variables $X = (X_1, X_2 ... X_p)$.

Situations that can be thought of in this way are very common. For example, the resource units can be items of prey, some of which are selected by predators. In that case appropriate X variables might indicate the size, colour and species of the units. Alternatively, the resource units might be plots of land, some of which are used by an organism. Appropriate X variables might then be percentages of different types of vegetation, the distance to water and the altitude.

In terms of estimation, there is a variety of situations that need to be considered within this general framework. Some important distinctions are:

(a) In some studies all the available resource units can be censused and classified as used or unused, but in other studies it is possible only to sample resource units. Different statistical models are required for these two types of study because in the first case errors in estimating the resource selection probability function only come about because resource selection is a stochastic process, but in the second case sampling errors are also involved. As will be seen later, it turns out that, with sample data, it is possible to estimate only the resource selection probability function multiplied by an arbitrary constant, unless information about sampling fractions is available.

(b) Some studies involve observing a single episode of selection but other studies involve several periods of selection, with more and more units being used as time accumulates. It is more straightforward to analyse data from the first type of study since there is no need to model the effect of increasing selection time.

(c) In some studies the resource units are characterized by the particular categories into which they fall, such as the type of habitat that they represent. In other studies each unit is either categorized in several ways or has several quantitative variables measured on it. The first type of study occurs rather commonly and it is therefore worthwhile to discuss it as a special case of particular interest.

2.2 EXAMPLES

It is useful at this point to discuss some examples of situations where the estimation of a resource selection probability function is a plausible approach to data analysis. All of these examples are considered more fully in later chapters. Here the intention is just to give a broad outline of the approaches that will be used.

2.2.1 Example 2.1 Habitat selection by moose

Neu *et al.* (1974) considered selection of habitat by moose (*Alces alces*) on a 33 200 acre site surrounding Little Sioux Burn in northeast Minnesota during the winter of 1971–72. They determined the proportion of the study area in four habitat categories (in burn, interior; in burn, edge; out of burn, edge; and out of burn, further) using a planimeter, and also, during aerial surveys classified 117 observations of groups of moose or moose tracks using the same categories. The results obtained are shown in Table 2.1.

Table 2.1 The occurrence of groups of moose or moose tracks on burned, unburned and peripheral portions of a study area surrounding Little Sioux Burn

Location	Proportion of total acreage	Observations of moose
In burn, interior	0.340	25
In burn, edge	0.101	22
Out of burn, edge	0.104	30
Out of burn, further	0.455	40
Total	1.000	117

In this study the resource units are not as clearly defined as is desirable, but are basically the 'points' in the study area where moose or moose tracks can potentially be observed. The question of interest is whether the comparison between the proportions of the study area in the four habitat categories and the corresponding proportions for the 117 observations indicates that the moose used the different types of habitat according to their availability, or selected in favour of one or more types. Since habitat availability was censused for the whole region, and use was sampled for the whole population of animals, this is an example of a design I study with sampling protocol A, as defined in sections 1.3 and 1.4.

In this example the resource selection probability function gives the probability that a resource unit in a particular category is used by the moose. However, it is clear that absolute probabilities of use cannot be estimated from the data in Table 2.1 since there is no way of knowing what proportion of resource units were used either overall, or for any of the four categories. Indeed, given the nature of the situation, it is difficult to say how these proportions should be defined.

What can be estimated is a resource selection function, which is the resource selection probability function multiplied by an arbitrary positive constant. In fact, in Chapter 4 it will be shown that the resource selection function is given by ratios of the observed to expected sample counts in different categories, or these ratios after they have been standardized in some way. Basically, this means that the function is given by the forage ratios that are defined in Table 1.1.

Thus in the situation where observations on resource use are considered to fall into several non-overlapping categories, the estimation of a resource selection function reduces to standard procedures. Or, to put this another way, the standard procedures can be justified on the basis of a general theory of resource selection functions.

2.2.2 Example 2.2 Habitat selection by pronghorn

As a second example, consider a study carried out by Ryder (1983) on winter habitat selection by pronghorn (*Antilocapra americana*) in the Red Rim area in south-central Wyoming. The study area consisted of blocks of alternating public and private land, and Ryder systematically sampled the public land to obtain 256 study plots of 4 ha each, covering 10% of the total public area. On each study plot Ryder recorded the presence or absence of antelope in the winters of 1980–81 and 1981–82, the density and average height of big sagebrush (*Artemisia tridentata*), black greasewood (*Sarcobatus vermiculatus*), Nuttall's saltbush (*Atriplex nuttalli*) and Douglas rabbitbrush (*Chrysothamnus viscidiflorus*), the slope, the distance to water, and the aspect. The data for the variables other than the vegetation heights are shown in Table 2.2. The vegetation height variables have been omitted here on the grounds that they are missing for plots on which the vegetation in question does not exist.

The aspect variable with values 1 to 4 cannot be used, as it stands, in a resource selection probability function since it implies, for example, that North/Northwest has four times as much 'aspect' as East/Northeast. However, as will be more fully discussed in example 5.1, this problem is easily overcome by converting the single variable shown in Table 2.2 into three 0–1 indicator variables. With this modification, Ryder's study plots can be thought of as a population of N = 256 available resource units, each of which has values for four vegetation density variables (X_1 to X_4), the slope (X_5), the distance to water (X_6), and three indicator variables for the aspect (X_7 to X_9). Since habitat availability and use is considered for the whole population of antelopes in the study region, this is an example of a design I study with census data in the terminology of sections 1.3 and 1.4.

Table 2.2 The presence and absence of pronghorn on 256 plots of public land in 1980–81 and 1981–82

	Use variables		Density of vegetation				Distance to		
Plot	80–81	81–82	Sage brush	Grease wood	Salt bush	Rabbit brush	Slope	Water	Aspect
1	0	0	9	6.8	2	0	0	25	4
2	0	1	18	0.6	1.6	0.6	5	150	3
3	0	0	8.4	0.8	0	12	45	150	4
4	0	0	3.2	0	0	4.2	65	375	2
5	0	1	12	0.2	0	0.6	5	375	3
6	1	1	7.8	2.6	10.4	0	5	150	3
7	0	0	5.4	0	0	7.8	55	625	1
8	1	0	10	3	0.2	0	5	150	1
9	1	0	12	0.2	0.6	2	5	875	3
10	1	1	12	0.2	4.6	0	15	375	3
11	0	0	0.6	0	3.4	0	75	625	2
12	1	1	7.6	0	3.4	4.4	5	25	1
13	0	0	4.2	0	0.2	0.4	45	1250	3
14	1	1	12	0.2	4.4	0.2	5	375	3
15	0	0	8.2	0	0	5.6	25	1250	3
16	0	1	4	0	0	0.4	25	1250	2
17	0	0	10	0	0	21	35	1250	4
18	1	0	4	0	0	11	15	1750	4
19	0	0	3.4	0	0	11	75	1250	2
20	0	0	6.4	0	0	6.4	15	875	4
21	1	0	4	0	0	5.4	5	1250	4
22	0	1	7.8	0	0	0.8	45	875	4
23	0	0	10	0	0	0.2	5	875	4
24	0	0	10	3.6	0.2	0.8	15	1250	4
25	1	0	3.8	0	0	0.4	5	375	1
26	0	0	1.2	0	3	0	45	875	3
27	0	0	2	0	1.2	0	5	625	4
28	1	0	5.8	0.2	9.4	0.2	5	875	3
29	1	1	7.4	0	0	4.6	5	375	1
30	1	0	7.2	0	0.4	0.2	5	875	3

Table 2.2   (cont.)

Plot	Use variables 80–81	81–82	Sage brush	Grease wood	Salt bush	Rabbit brush	Slope	Water	Aspect
31	0	0	4	0	0	6	55	875	3
32	1	0	18	0	0	17	5	375	4
33	0	1	1.6	0.2	7.8	0.6	5	875	3
34	0	1	3.4	4.8	4.2	2	0	25	4
35	0	1	0.8	0	0	0.6	35	875	1
36	0	1	5.9	0	8.2	0.2	5	875	1
37	1	0	0.2	0.2	2.9	5.4	15	375	3
38	1	0	12	0	0	1.2	5	625	3
39	0	0	9	0	0	7	65	625	3
40	0	0	2	0	2.4	4.2	15	625	4
41	1	1	9.2	0	0	5.6	85	375	1
42	0	1	6.2	0	1.2	0.3	25	875	4
43	1	1	0.2	3	11	0	0	25	4
44	0	0	7.4	0	0	3.8	15	625	3
45	0	1	3.2	0	4	0.2	25	150	2
46	0	0	2.2	3.8	19.2	0	45	150	4
47	0	0	9.6	0	1.6	0.2	15	25	2
48	0	0	13	0	0	28	15	625	4
49	0	1	11	2	3.4	0.4	15	625	4
50	0	1	2.6	0	0	1	35	875	4
51	1	1	4.8	6.8	8.4	0	0	25	4
52	0	0	6.6	0.6	2.8	1	0	375	4
53	0	1	16	0.2	0	12	15	875	4
54	0	1	12	0	0	18	55	1750	4
55	0	0	3	0	0	0.6	15	1750	3
56	0	0	0.7	0	0	0	45	1250	4
57	0	0	3	0.2	3.6	0.2	75	875	4
58	0	0	2.6	0	0	2	0	1250	4
59	0	1	5.4	1	4.2	1	55	1250	2
60	1	0	5.6	1	11.4	0	5	875	3
61	1	0	3.8	10	18.2	0	0	150	4
62	1	1	0	0	0	0	25	1250	4
63	0	0	0.4	0	0	0	15	1250	2
64	0	1	1.4	0	0	0.2	15	1250	3
65	1	0	0.4	0	8.3	0	5	375	2
66	0	1	4.8	0	0	0	5	875	4
67	0	0	5.6	0	2.8	0.2	25	1250	3
68	0	0	0.2	0.2	18.4	0	5	625	3
69	0	1	3.9	0	0	0	0	25	4
70	0	1	10	0	0	5.6	25	875	4
71	0	0	4.2	0	0	0.2	25	875	4

Table 2.2   (cont.)

Plot	Use variables		Density of vegetation				Distance to		
	80–81	81–82	Sage brush	Grease wood	Salt bush	Rabbit brush	Slope	Water	Aspect
72	1	0	0.2	0	15.6	0	5	875	3
73	0	1	0.2	0.2	0	0	0	150	4
74	1	0	5	0	0	0	15	625	3
75	0	0	5.4	0.2	0	1	55	1250	4
76	1	1	3.4	1.2	31.4	0	25	625	1
77	0	1	10	0	0.2	0.8	15	375	4
78	0	0	0.4	0	11.2	3.6	5	625	4
79	1	1	4.8	0.2	0.2	4.2	25	625	3
80	0	1	7.2	4.4	6.8	2.4	0	25	1
81	1	0	6.6	1.2	17	1	85	150	4
82	1	1	1.6	0	0	1	45	875	4
83	1	1	4.2	4.4	12.4	0	5	150	1
84	0	0	5	0	0	2.2	25	1750	3
85	1	1	3.4	0	17	0	15	875	3
86	0	0	0.8	0	11.8	0	85	1250	4
87	0	0	0.8	0	0	1	15	875	3
88	0	0	8	0	0	3	35	2250	1
89	0	0	0.2	0	5.2	0.2	5	1250	1
90	0	0	10	0	0	4	15	1250	4
91	0	0	0.2	0	45.4	0.2	15	375	4
92	0	0	4.2	0	0	1.4	15	1750	1
93	0	0	7	0	0	1.6	15	875	1
94	0	0	2.4	0.4	0.2	6.6	35	1250	1
95	0	0	10	0	0.6	6.4	55	875	4
96	0	0	7.8	0	0	0.6	15	2250	4
97	1	1	12	0	0.4	4.4	25	1250	1
98	0	0	3.2	0	0	0.8	25	1250	4
99	1	1	1	0.2	1.6	0.8	15	375	3
100	0	1	5.4	0	22.4	1	5	1750	4
101	0	1	11	0	2.2	2.4	0	1250	4
102	0	1	0.8	0	0	4.4	15	875	4
103	0	0	3.6	0	6.2	3.2	5	625	2
104	0	1	5.4	0	0	2.2	25	2250	3
105	1	1	9	0	0	1.8	5	1750	4
106	0	0	2	0.2	26.4	0	5	875	4
107	0	0	0.6	0	0	10	0	150	4
108	0	1	4.2	0	2.2	1.2	15	1750	4
109	0	1	1.8	0	2	0.2	5	1250	4
110	1	1	4.4	2	1.4	9.6	25	375	4
111	1	0	0.2	0	0	0.6	15	375	1
112	1	1	0.2	0	7	0	15	2250	3

Table 2.2 (cont.)

Plot	Use variables 80–81	Use variables 81–82	Sage brush	Grease wood	Salt bush	Rabbit brush	Slope	Water	Aspect
113	1	0	3.8	2.8	0.6	1.6	15	1750	4
114	1	1	0.2	0.2	26.6	0	0	625	4
115	1	1	0.4	0	0	0	0	25	4
116	0	0	7.2	0	0	0.4	35	2750	4
117	1	1	9.4	0	0	0.2	15	1750	4
118	0	0	0.2	0	24.6	0.2	15	875	4
119	0	1	4	0	0	11	35	375	4
120	0	0	3.6	0	0	1.4	45	2750	2
121	0	0	2.4	0	0	7.6	15	2250	4
122	1	1	5	0	0.4	1.4	25	875	4
123	0	1	1.4	0.2	11.4	0	0	375	4
124	0	0	8.2	0	0	16	5	2250	4
125	0	0	7.4	0	0	3.8	15	1750	2
126	0	0	2	1.4	18.4	0.2	0	375	4
127	0	0	0.4	0	9	1	15	625	4
128	0	0	4.4	0	0	2.6	5	2750	4
129	1	1	0.2	0	1.2	0.2	25	2250	1
130	0	0	3.6	0.6	5.4	0.2	5	625	4
131	0	0	2.4	0.2	5.2	0	0	875	4
132	1	0	10	0	0	11	15	2250	4
133	0	0	8.4	0	5.8	0.2	25	1750	4
134	0	0	0	0	0.4	0	45	25	3
135	0	1	4	1.2	17	0	0	625	4
136	1	0	4.8	0	0	7.2	25	2750	3
137	0	0	1.4	0	1.2	1.4	15	2250	4
138	1	1	0	0	0	0	35	375	4
139	0	0	2.4	0.4	20.4	0.2	5	625	1
140	0	0	3.4	0	13.4	0.2	15	2750	4
141	0	0	3	0.2	0	0.4	15	1750	4
142	1	0	5	0	0.2	6.6	15	375	4
143	0	1	0	0	5	0.4	5	150	2
144	0	0	8.4	0	0	3	5	2750	4
145	1	0	7.6	1.4	0.4	0	25	2750	4
146	0	1	0.2	0.2	11.6	0	15	375	4
147	0	0	2.8	0.2	0.2	11	0	625	4
148	1	0	3.4	0	0	2.2	0	2750	4
149	0	1	2.4	0	0	2.2	35	2750	3
150	1	1	1	0.8	16	0.2	15	1750	2
151	1	1	2.2	2	25.2	0	0	625	4
152	1	1	0.4	8	2.2	0	0	375	4
153	0	0	1.6	0	0	4.4	45	2750	2

Table 2.2   (cont.)

Plot	Use variables		Density of vegetation				Distance to		
	80–81	81–82	Sage brush	Grease wood	Salt bush	Rabbit brush	Slope	Water	Aspect
154	0	0	7.6	0	0	1.2	15	2750	4
155	0	1	0.4	0	5	5.7	15	875	4
156	0	0	0.6	1.4	21.2	0	5	625	4
157	0	0	5.6	7.6	14.2	0	0	25	4
158	0	0	14	0	0	0.2	15	2750	4
159	0	1	3.6	0	2.8	3.4	15	2750	4
160	1	1	2.2	5.8	8.2	0	0	625	4
161	0	0	6.6	0	1.8	2.4	45	2750	4
162	0	0	11	0	0.4	3.6	15	2750	3
163	1	1	3	0.2	17.2	0.2	15	875	2
164	0	0	3.4	0.2	3.4	5.8	35	375	4
165	0	1	12	7.2	0.8	0.6	0	25	4
166	0	0	10	0	0	1.6	25	2750	3
167	1	0	3.8	0.2	1.4	1	15	2750	1
168	0	0	0.4	1.8	18	0	0	150	4
169	1	1	12	0	0	7	35	2750	1
170	0	1	1	0	0	0	15	2750	4
171	1	0	0.4	6.1	0.4	0	0	875	4
172	1	0	1.2	8.2	18.2	0	0	625	4
173	0	1	6	0	4.4	2	5	2750	4
174	0	0	6.2	0	0.2	1.8	25	2250	1
175	0	0	3	0.2	1.4	14	15	1750	3
176	1	1	0.4	0.6	19.6	0	0	625	4
177	0	1	2	0.2	10.6	0.2	0	25	4
178	0	0	1.2	0	8.8	0.2	5	2250	2
179	0	1	0	0	0.4	0	45	1750	4
180	0	0	0.4	0	6.8	0	0	1250	4
181	1	1	0.2	0.2	24.6	0	0	875	4
182	0	0	10	0	0	0.6	15	2750	3
183	0	0	7.2	0.2	0.6	0.4	25	1250	4
184	0	1	6.2	0.8	1.6	3	25	1750	4
185	1	1	0.2	0.2	0.2	0.2	0	875	4
186	0	0	1.8	1.2	1.8	0	0	25	4
187	0	0	7.2	0	0	8.4	55	2750	3
188	0	0	4.8	0	0	1	15	2750	2
189	0	1	3.2	0	0.4	3.6	15	2750	1
190	1	0	3.4	0.2	3	1.2	5	1750	4
191	0	0	0.2	0	8.8	2.6	5	1750	2
192	0	1	4	6	2	0.2	0	875	4
193	0	0	4	2	2	0.2	0	150	4
194	0	0	6	0	0	11	15	2750	3

Table 2.2   (cont.)

Plot	Use variables		Density of vegetation				Distance to		
	80–81	81–82	Sage brush	Grease wood	Salt bush	Rabbit brush	Slope	Water	Aspect
195	0	0	0.6	0.2	26.8	0	5	1250	4
196	0	0	1	0	6.2	3.8	0	1250	4
197	0	0	2	4	0.2	3.8	0	375	4
198	0	1	0.2	0	6.4	2	0	150	4
199	0	1	2.6	0	0	2.6	0	2750	4
200	1	1	2.8	0	0	2.6	25	2750	4
201	0	1	6.1	0	1.8	0.4	15	2250	2
202	0	1	3	0.8	5	1.2	15	875	4
203	0	0	6.2	0.6	0	0	15	1250	4
204	0	0	4	2.6	8	4	0	875	4
205	0	0	6	3	2	6	0	25	4
206	0	1	0.8	0	9.4	6	15	2750	1
207	0	1	0.2	0.2	19	0.2	5	875	4
208	0	0	1.8	0.2	13.6	0.8	35	1250	4
209	0	0	8	12	5	8	0	375	4
210	0	0	0.2	0.2	25.6	0	0	25	4
211	0	0	2.2	0	0	2	15	2750	2
212	1	0	6.6	0	0	11	5	2750	4
213	0	0	2.6	2.2	6.6	0.6	35	1750	3
214	1	0	6.6	2.2	3.4	0.4	25	875	2
215	1	0	0	0	6.4	0	15	875	4
216	0	0	1	3	0.2	0.6	0	625	4
217	0	0	0.4	0.4	4.8	0	0	375	4
218	0	0	9.2	0.2	0.8	7.8	5	2750	1
219	0	0	3.2	0.4	4	2	5	625	4
220	0	0	1.4	0.2	8	0.2	15	1250	4
221	0	0	6.2	0.2	0.2	0.2	0	375	4
222	0	0	6	0	0	3.2	15	2750	2
223	0	1	6.4	0	1.4	3.4	5	2750	4
224	0	0	1	4.2	15.2	0	5	1750	4
225	1	1	0.2	0.2	17	0	15	375	1
226	1	0	1	0	0.4	0.7	15	625	4
227	0	0	0	0	0	0	0	875	4
228	1	0	3.2	0	0	2.2	0	25	4
229	0	0	0.2	1.4	1.4	0.2	25	1750	2
230	0	0	2	2.6	12	0.2	5	1250	1
231	1	1	1.4	0	4.8	1	45	2250	2
232	0	0	0	0	1.4	0	35	1750	4
233	0	1	0	0	0.4	0	0	1750	4
234	0	1	0	11	14.6	0	0	1250	4
235	0	1	1.4	0	0.2	0.8	85	2250	3

Table 2.2   (cont.)

Plot	Use variables		Density of vegetation				Distance to		
	80–81	81–82	Sage brush	Grease wood	Salt bush	Rabbit brush	Slope	Water	Aspect
236	1	1	0.2	0.2	23.8	0.2	35	1750	2
237	0	0	0	0	8	0	15	2250	4
238	0	0	0.6	0.2	11.4	1.2	15	1250	3
239	0	0	0	0	0	0	5	2750	3
240	0	0	2.4	0.2	22.2	0.6	5	1750	4
241	0	0	1.6	0	0.2	0.8	15	2750	2
242	0	0	2.2	0	3	2.2	15	2250	3
243	1	0	1.1	0	15.1	0	25	1750	4
244	0	0	3.2	0	7.8	3.4	5	1750	4
245	0	0	0	0	0	0	15	2750	4
246	0	1	1.8	0.2	0	1.6	5	2750	2
247	0	0	4.2	0	10	0.2	65	2750	2
248	0	0	0.8	0	12.6	0.2	5	2750	4
249	0	0	5.2	0	0.2	4.2	5	2750	2
250	0	0	0.2	0	11.8	0	15	2750	2
251	1	0	0	0	2	0	85	2750	4
252	0	1	6.4	0	2.8	5.2	5	2750	4
253	0	0	3.4	0	0	8.2	75	2750	4
254	0	0	1.8	0	14.4	0	5	2750	4
255	0	0	0.4	0	9.7	0.4	5	2750	4
256	0	0	0.4	10	5	0	75	2750	4

The variables in order from left to right are: the plot number; use in 1980–81 (1 for use, 0 for no use); use in 1981–82; sagebrush density (thousands/ha); greasewood density (thousands/ha); saltbush density (thousands/ha); rabbitbrush density (thousands/ha); mean slope of plot (degrees below horizontal); distance from centre of plot to water (m); and the aspect (1 for East/Northeast, 2 for South/Southeast, 3 for West/Southwest, and 4 for North/Northwest).

In this example the resource selection probability function gives the probability of a study plot being used as a function of X_1 to X_9. However, since the presence and absence of pronghorn is recorded in 1980–81 and 1981–82 there are several different ways to define 'use', with correspondingly different resource selection probability functions. Thus, plots can be considered to be 'used' if pronghorn are recorded in either winter, or, alternatively, if pronghorn are recorded in both winters. Other possibilities are to think of the two winters as replicates of the selection process, or to think of each plot as being observed after one and two years of selection, with a plot being considered as 'used' when pronghorn are first observed.

One of the most straightforward ways of estimating a resource selection probability function involves assuming that the probability of observing pronghorn in one winter, on a plot with values $x = (x_1, x_2, \ldots, x_9)$ for the nine X variables, is given by the logistic regression equation

$$w^*(x) = \{\exp (\beta_0 + \beta_1 x_1 + \ldots + \beta_9 x_9)\} / \{1 + \exp (\beta_0 + \beta_1 x_1 + \ldots + \beta_9 x_9)\}$$

where β_0 to β_9 are unknown parameters to be estimated. This is the approach adopted in example 5.1, with the estimation of the β parameters being carried out using the logistic regression option of the SOLO computer package (BMDP, 1988).

2.2.3 Example 2.3 Selection of snails by birds

An experiment described by Bantock *et al.* (1976) on the selection of *Cepaea nemoralis* and *C. hortensis* snails by the song thrush (*Turdus ericetorum*) is an example of a situation where it is necessary to consider the effect of different periods of selection time. The experimental site was a disused camellia-house covering 39 m^2 with ground vegetation of nettles that was almost empty of snails before the experiment began on 29 June, 1972. On that date, 498 yellow five-banded (Y5H) *C. hortensis*, 499 yellow five-banded (Y5N) *C. nemoralis*, and 877 yellow mid-banded (Y3N) *C. nemoralis* snails were released into the area. Shells were uniquely marked so that the survivors could be determined from censuses taken at various times after the population had been set up. The nearest natural population of *Cepaea* was 100 m from the experimental site. Thrush predation occurred there, but stopped when the experimental snails were released.

Simplified results from the experiment are shown in Table 2.3. Only five-banded snails are considered since extra mid-banded *C. nemoralis* were added to the population after 29 June. The methods that we propose for estimating a resource selection probability function using the simplified data can be extended to take into account this type of experimental procedure by thinking of each augmentation of the prey population as the start of a 'new' experiment, but we consider that it is better not to introduce this complication into an illustrative example.

In Table 2.3 different types of snail are defined in terms of two X variables, where X_1 is a species indicator which is 1 for *C. nemoralis* and 0 for *C. hortensis*, and X_2 is the maximum shell diameter in units of 0.3 mm over 14.3 mm. The table shows that there was a population of $N = 997$ available resource units (snails) at the start of the experiment, each with values for $p = 2$ characterizing variables, and the survivors are known after 6, 12 and 22 days of selection.

Clearly the time element cannot be ignored in this example, so that the resource selection probability function has to give the probability of a snail being eaten after t days of selection. There are various ways of estimating such a function, one of which is discussed in example 6.1, where it is assumed that the probability of use by day t is given by

$$w^*(x,t) = 1 - \exp \{-\exp (\beta_0 + \beta_1 x_1 + \beta_2 x_2)t\}.$$

Table 2.3 Eaten and uneaten snails in an experimental population of yellow five-banded *Cepaea nemoralis* and *C. hortensis*

Shell diameter	Species (X_1)	Coded size (X_2)	Snails eaten between days 0–6	6–12	12–22	Left on day 22
14.6	0	1	0	0	0	2
14.9	0	2	1	0	0	2
15.2	0	3	0	0	0	2
15.5	0	4	0	1	1	3
15.8	0	5	0	1	0	4
16.1	0	6	0	3	3	6
16.4	0	7	2	2	3	12
16.7	0	8	3	1	9	23
17.0	0	9	7	6	9	20
17.3	0	10	5	8	8	35
17.6	0	11	6	9	13	29
17.9	0	12	6	10	11	32
18.2	0	13	8	9	10	35
18.5	0	14	7	12	11	26
18.8	0	15	4	9	6	17
19.1	0	16	1	3	1	11
19.4	0	17	5	1	1	7
19.7	0	18	1	0	2	5
20.0	0	19	0	1	1	1
20.3	0	20	0	0	0	4
20.9	0	22	0	0	0	1
16.7	1	8	0	0	0	1
17.3	1	10	0	0	1	0
17.6	1	11	0	0	0	1
17.9	1	12	2	1	0	2
18.2	1	13	4	2	3	0
18.5	1	14	2	2	1	9
18.8	1	15	5	5	5	7
19.1	1	16	3	8	8	11
19.4	1	17	10	14	15	30
19.7	1	18	19	18	12	19
20.0	1	19	21	17	4	16
20.3	1	20	18	17	11	13
20.6	1	21	13	8	10	21
20.9	1	22	13	10	8	13
21.2	1	23	5	7	2	12
21.5	1	24	7	1	3	4
21.8	1	25	3	2	2	5
22.1	1	26	1	2	0	3
22.4	1	27	0	1	2	0
22.7	1	28	2	0	0	0
23.0	1	29	1	0	0	0

The species indicator variable X_1 is 0 for *C. hortensis* and 1 for *C. nemoralis* and the coded size variable is the shell diameter in units of 0.3 mm over 14.3 mm, rounded to the nearest unit.

This corresponds to assuming that the probability of not being used is given by the proportional hazards function

$$\phi^*(x,t) = \exp\{-\exp(\beta_0 + \beta_1 x_1 + \beta_2 x_2)t\},$$

which is often used for analysing survival data.

Since the available snails are assumed to be the same for all of the birds selecting them, and no information is available on the choices of individual birds, this is another example of a design I study in the terminology of section 1.3, with census data.

2.2.4 Example 2.4 Nest selection by fernbirds

An example where only samples of available and used resource units were taken is provided by a study of nest selection by fernbirds (*Bowdleria puncta*) in Otago, New Zealand, that is described by Harris (1986). Harris measured nine variables on 24 nest sites found during the 1982–83 and 1983–84 seasons and found comparative available sites by choosing 25 random points in the study area and locating a polystyrene model nest at the centre of the nearest clump of vegetation. Harris concluded that only three of the nine variables that he measured showed important differences between the two samples. The data for these variables (the canopy height, the distance from the outer edge of the nest to the nearest outer surface of the clump of vegetation in which the nest is situated, and the perimeter of the clump of vegetation) are shown in Table 2.4.

In this example the total number of available resource units (potential nest sites) is unknown, but clearly very large, and there are $p = 3$ variables measured on each sampled resource unit. From the data available there is no way of estimating the absolute probability of a potential nest site being used. However, it is possible to estimate a resource selection function which is this probability multiplied by an arbitrary constant.

One approach to estimating the resource selection function involves assuming that it takes the form

$$w(x) = \exp(\beta_0 + \beta_1 x_1 + \beta_2 x_2 + \beta_3 x_3).$$

This then leads to a model for the sample data, which can be estimated by any standard computer program for log-linear modelling, as is discussed in example 7.1. Alternatively, exactly the same model can be estimated by using logistic regression, as is discussed in example 8.1.

Harris' study has design I with sampling protocol A in the terminology of sections 1.3 and 1.4 since availability was measured by sampling potential nest sites over the entire study region, and there is the implicit assumption that the probability of a potential site being used was approximately the same for all fernbirds.

Table 2.4 Comparison of variables measured on fernbird nest sites and randomly located sites in the same area

	Nest sites			Available sites		
	Canopy height (m)	Distance to edge (cm)	Perimeter of clump (m)	Canopy height (m)	Distance to edge (cm)	Perimeter of clump (m)
	1.20	14.0	8.90	0.47	13.5	3.17
	0.58	25.0	4.34	0.62	8.0	3.23
	0.74	14.0	2.30	0.75	19.0	2.44
	0.70	12.0	5.16	0.52	5.0	1.56
	1.36	14.5	2.92	0.73	8.0	2.28
	0.78	17.0	3.30	0.62	16.0	3.16
	0.45	15.0	3.17	0.60	17.0	2.78
	0.78	15.0	4.81	0.26	4.5	3.07
	0.63	16.0	2.40	0.46	15.0	3.84
	0.75	12.0	3.74	0.28	12.0	3.33
	0.55	12.0	4.86	0.53	11.0	2.80
	0.45	20.0	2.88	0.42	17.0	2.92
	1.56	16.0	4.90	0.47	20.0	4.40
	0.85	23.0	4.65	0.50	13.0	3.86
	0.58	12.0	4.02	0.54	16.0	3.48
	0.75	13.0	4.54	0.56	18.0	2.36
	0.55	18.0	3.22	0.32	7.0	3.08
	0.56	18.0	3.08	0.62	16.0	5.07
	0.57	19.5	4.43	0.39	15.0	2.02
	0.41	16.0	3.48	0.56	8.0	1.81
	0.65	18.0	4.50	0.27	9.0	2.05
	0.78	14.0	2.96	0.42	11.0	1.74
	0.64	18.0	5.25	0.70	13.0	2.85
	0.71	18.0	3.07	0.26	14.0	3.64
				0.34	9.5	2.40
Mean	0.73	16.3	4.04	0.49	12.6	2.93
SD	0.28	3.4	1.37	0.15	4.4	0.84

Example 2.5 Selection of corixids by minnows

Although most resource selection studies based on sample data consider only a single selection period, there are cases where a population of resource units has been sampled at several times while selection is proceeding. The situation is then similar to that in example 2.3 but with samples taken instead of censuses.

An example is a study of the use of corixids as food by minnows (*Phoxinus phoxinus*). Popham (1944) sampled the corixids in a pond every day from 13 to 19 September, 1942, introduced 50 minnows on the evening of 19 September, and then sampled again every day from 22 to 28 September. This resulted in the sample counts shown in Table 2.5 for three corixid species, with shades of grey classified as light, medium, and dark. The samples taken before the introduction of minnows have been lumped together to form a single 'available sample' since they have similar proportions for the species and colours of corixids.

Table 2.5   Results from sampling a corixid population before and after minnows were introduced into a pond on 19 September, 1942

Corixid species	Shade of grey	Available samples	Samples of live corixids on September						
			22	23	24	25	26	27	28
Sigara venusta	Light	120	5	5	5	6	2	6	3
	Medium	726	102	110	131	120	105	157	134
	Dark	225	25	58	22	22	16	20	17
Sigara praeusta	Light	25	2	0	2	1	0	1	0
	Medium	33	6	2	6	3	5	4	5
	Dark	6	1	1	0	0	0	0	1
Sigara distincta	Light	15	0	0	1	0	0	0	0
	Medium	39	4	3	3	4	3	4	5
	Dark	6	0	0	0	0	0	0	0

Popham argued that any changes in the relative proportions of the nine types of corixid after 19 September were largely due to predation by the minnows since the effects of immigration and emigration of corixids were negligible, and newly formed adults were entering the population at a low rate. Therefore this can be thought of as a situation where there are eight samples of unused resource units (corixids), taken after selection times of 0, 3, 4,..., 9 days.

There are nine types of resource unit in this example, which differ because of their species and their colour. Although these are qualitative rather than quantitative differences it is still possible to use X variables to describe the units, and hence fit this example within the framework of the present chapter. Essentially, all that is necessary is to set up appropriate 0–1 indicator variables, as discussed more fully in example 7.3.

Because of the nature of the data, with only samples of unused resource units being available, it turns out that it is not possible to estimate a resource selection probability function, or even this function multiplied by an arbitrary constant. However, what can be done is to assume that the probability of a unit *not* being used by time t takes the form

$$\phi^*(x,t) = \exp \{(\beta_0 + \beta_1 x_1 + \ldots + \beta_p x_p)t\},$$

where the exponential parameter is necessarily negative. It then becomes possible to describe the sample counts of different types of corixid using a log-linear model, and hence estimate the parameters β_1 to β_p using a suitable computer program.

Although β_0 cannot be estimated, it is possible to estimate

$$\phi(x,t) = \exp\{(\beta_1 x_1 + \beta_2 x_2 \ldots + \beta_p x_p)t\},$$

which gives the probability of not being used by time t multiplied by an arbitrary constant. The estimated function can then be used to rank the types of corixid in order of the probability of not being used, which is the reverse order to that for the probability of use. In this way, the selection of corixids can be studied.

Because the availability of corixids is assumed to be the same for all minnows, and there is no information about the choice of corixids by individual minnows, this is another example of a design I study as defined in section 1.3, but with sampling protocol B as defined in section 1.4 since unused resource units are compared with those available.

2.3 SAMPLE DESIGNS

Although all of the examples just considered have designs that are classified as being of design I in the terminology of Chapter 1, this should not be taken to mean that this is inevitable when a resource selection probability function or a resource selection function can be estimated. The methods proposed in this and the following chapters can be used with design II studies where the resources used by individual animals are measured but the availability of resources is measured at the population level. All that needs to be done is to estimate a separate selection function for each animal if this is necessary. Similarly, with design III studies, where resource use and availability is measured for several animals, it is possible to estimate a separate selection function for each animal if necessary. The choice of examples in this chapter just reflects the fact that design I studies have been most common in the past.

2.4 ASSUMPTIONS

There are some general assumptions that are involved in the estimation of resource selection functions that are worth listing at this point since they will generally be implicitly assumed to hold for all examples. These assumptions are that:

(a) the distributions of the measured X variables for the available resource units and the resource selection probability function do not change during the study period;

(b) the population of resource units available to the organisms has been correctly identified;

(c) the subpopulations of used and unused resource units have been correctly identified;

(d) the X variables which actually influence the probability of selection have been correctly identified and measured;

(e) organisms have free and equal access to all available resource units; and

(f) when studies involve the sampling of resource units, these units are sampled randomly and independently.

These assumptions will now be briefly reviewed in the order just given.

The requirement that the distributions of the X variables for the available resource units and the resource selection probability function do not change during the study period is difficult to satisfy with many studies. With sample data the population of available resource units may change without the investigator realizing that this has happened. The resource selection probability function may change with the season, the weather, the distribution of remaining resources, etc. In all cases, inferences are made with respect to 'averages' over the period when used and unused resource units are collected. If populations are not changing rapidly then this should not be a problem but generally the selection times should be kept as short as possible. In rapidly changing populations, an attempt should be made to obtain several 'snapshots' of resource selection over time.

Identification of the population of resource units that are available to the organisms, and the variables which influence the probability of selection, are probably the most crucial and most difficult aspects of the study design. Not much advice can be given here to help the researcher in specific cases. However, as noted by Manly (1985, p. 172) and Rexstad et al. (1988), the use of stepwise procedures to identify important variables from a larger set of potentially important variables may to give misleading results and should be used with caution.

The assumption that organisms have unrestricted and equal access to all the available resource units is most easily justified when the subpopulation of used units is small relative to the population of available units. Of course, changes in the density of animals or in the availability of resource units may change the underlying selection strategies and the selection function. Thus, statistical inferences are made with respect to the specific densities present in the study area.

If resource units are not sampled randomly and independently then the estimates of the coefficients of a resource selection probability function may still be meaningful but standard errors may not reflect the true variation in the populations. In many situations it is difficult to design studies so that the individual resource units are the basic sampling units and hence avoid pseudoreplication (Hurlbert 1984). For example, relocations of radio-tagged animals may be recorded at a series of points in time. Care must then be taken to ensure that the time interval between recordings is sufficient to assume that identifications of used habitat points are independent events if standard model based analyses of the data are to be conducted. Another complicating factor is that different collection methods are often required to obtain used and unused units.

Given these problems, one approach is to estimate a separate resource selection function for several independent replications of batches of dependent units. Alternatively, one might consider the selection of individual animals as independent events, and estimate a separate selection function for each animal by randomly sampling the units available and the units used by each animal. This may be the only reasonable approach for study of food and habitat use by highly territorial animals.

Another – less satisfactory – way of handling pseudoreplication that can be used with some of the models discussed in the following chapters involves estimating a heterogeneity factor as the ratio of an observed chi-squared goodness-of-fit statistic to the value that is expected from random sampling. The variances of all parameter estimates can then be multiplied by the heterogeneity factor to adjust them for the 'extraneous' variance. A problem with this approach is that it depends on the assumption that the variances of the observed data values are all inflated by the same amount, which may well not be true. Also, the method can be used only in cases where the counts of different types of resource unit are large enough to permit a valid chi-squared goodness-of-fit statistic to be calculated.

This method is discussed further in example 7.2 where it is suggested as a possible means of allowing for the fact that samples of eaten *Daphnia publicaria* come from cluster sampling of yellow perch stomachs instead of the random sampling of the entire population of predated *D. publicaria*.

3 Statistical modelling procedures

The statistical procedures that will be used in the following chapters often involve fitting several models to each set of data and deciding which is the simplest model that accounts adequately for the observed variation. The chosen model is then used to assess the amount of resource selection. Within this model-based framework there is a variety of inference procedures that are useful. These procedures are reviewed in this chapter.

3.1 SIMPLE SAMPLE COMPARISONS

Resource selection studies involve the comparison between samples or censuses of used, unused and available resource units. Therefore, at an early stage the analysis of data should involve the comparison of the distributions of X variables for the samples being compared. This will provide an indication of the differences, if any, that are present, and highlight any unusual aspects of the distributions. Graphical comparisons are always useful, and these can be supplemented if desired by parametric or non-parametric tests to compare means and variances. Univariate tests can be carried out on individual variables, and multivariate tests can be carried out on several variables simultaneously. Chi-squared tests can be used to compare samples in terms of the proportions of units in different categories.

If formal tests are carried out then some consideration should be given to the assumptions that are required to make these tests valid. The question of interest is whether the division of resource units into the two groups of used and unused ones has been made at random with respect to the X variables that are used to characterize the units. This suggests that randomization tests as discussed by Manly (1991) should be considered instead of tests that involve model based assumptions. However, since tests for significant differences between used and unused resource units can be carried out as part of the process for estimating the resource selection probability function it is simpler to avoid making any formal tests during the initial comparisons between distributions for used and unused units.

3.2 TESTING COEFFICIENTS OF INDIVIDUAL X VARIABLES

The estimation of resource selection probability functions and related functions amounts to estimating the coefficients of the X variables together with approximate values for the standard errors associated with the estimated coefficients. An obvious step in the analysis therefore involves checking to see whether the estimate $\hat{\beta}_i$ of β_i, the coefficient of x_i, is significantly different from zero. To this end, the hypothesis that $\beta_i = 0$ can be tested by comparing $\hat{\beta}_i/se(\hat{\beta}_i)$ with critical values from the standard normal distribution, where $se(\hat{\beta}_i)$ is the standard error of $\hat{\beta}_i$. For example $|\hat{\beta}_i/se(\hat{\beta}_i)| > 1.96$ implies that $\hat{\beta}_i$ is significantly different from zero at about the 5% level on a two-sided test. Also, an approximate 95% confidence interval for β_i is $\hat{\beta}_i - 1.96.se(\hat{\beta}_i)$ to $\hat{\beta}_i + 1.96.se(\hat{\beta}_i)$.

The theoretical justification for using these tests and confidence intervals comes from the fact that all the parametric models that are considered in this book for resource selection functions can be estimated by maximum likelihood, and the estimators of parameters are normally distributed for large sets of data. Precisely what constitutes a 'large' set of data depends on the circumstances, and in practice the normal approximation is best regarded with some reservations unless the data being analysed are from 100 or more resource units.

3.3 LOG-LIKELIHOOD CHI-SQUARED TESTS

When a parametric model for a resource selection probability or related function is correct, and the parameters of the model are estimated by maximum likelihood, then, under certain circumstances, minus twice the maximized log-likelihood approximately follows a chi-squared distribution with the degrees of freedom being the number of observations minus the number of estimated parameters. It follows that this chi-squared statistic, which will be denoted by X_L^2, may be used for testing the goodness-of-fit of an estimated model by comparing its value with the percentage points of the chi-squared distribution. Basically, if X_L^2 is not significantly large then a model can be considered to be reasonable.

The precise conditions for using this chi-squared test depend on the circumstances. However, in many cases it is a case of most of the expected counts of resource units in different categories being large, where the meaning of 'large' is the same as for other chi-squared goodness-of-fit tests. Hence the comparison of X_L^2 values with critical values from the chi-squared distribution is reasonable if most of these counts are five or more.

Another useful application of the chi-squared distribution comes from the result that if the p_1 parameters of one model are a subset of the p_2 parameters of a second model, and the first model is in fact correct, then the goodness-of-fit statistic for the first model minus the goodness-of-fit statistic for the second model, $D = X_{L1}^2 - X_{L2}^2$, will approximately be a random value from the chi-squared distribution with p_2-p_1 degrees of freedom. Therefore, if D is significantly large when compared with critical values of the chi-squared distribution then this indicates that the second model provides a significant improvement over the first model in fitting the data.

Like the chi-squared goodness-of-fit test, the use of the D statistic relies on the data set being 'large'. However, there is some evidence that the chi-squared approximation for D is more robust than the approximation for X_L^2 (McCullagh and Nelder, 1989, p. 119). Therefore tests for seeing whether one model gives a significantly better fit than a simpler model can be used with some confidence even with quite small sets of data.

A particular use of the D statistic that is important is the comparison of the fit of the model for which all resource units are used with equal probability (the 'no selection' model), with the fit of a model where the probability of a unit being used is a function of the X variables that are measured on the unit. If this comparison yields a significantly high D value then there is evidence of selection related to at least one of the X variables being considered.

The log-likelihood chi-squared statistic for a fitted model is sometimes described as the 'deviance', and regarded as being analogous to the residual sum of squares for regression and analysis of variance models. This leads to the idea of presenting the comparison between the fits of different models in terms of an analysis of deviance table that is similar to an analysis of variance table. Several tables of this type are provided in Chapters 5 to 9.

3.4 ANALYSIS OF RESIDUALS

The analysis of residuals (differences between observed and expected data values) is useful for highlighting any anomalous observations in a set of data, and seeing whether there are any patterns in the discrepancies between the model and the data that require further study. Hence, whenever possible this type of analysis should be carried out before accepting that a particular model is reasonable.

The residuals are easiest to interpret when they have been standardized so that, for the correct model, the mean should be approximately zero and the variance approximately one. If, in addition, they are approximately normally distributed then it is desirable to find most standardized residuals in the range from –2 to +2, and almost all of them within the range from –3 to +3.

The most obvious standardization involves dividing differences between observed and expected data values by the standard deviations of these differences. The nature of this standardization depends on the circumstances, but in the cases considered in the following chapters there are only two possibilities to be considered:

(i) If the ith data value O_i follows a binomial distribution with mean $A_i \Theta_i$ and variance $A_i \Theta_i (1-\Theta_i)$ then standardized residuals are

$$R_i = (O_i - A_i \hat{\Theta}_i)/\sqrt{\{A_i \hat{\Theta}_i (1-\hat{\Theta}_i)\}},$$

where $\hat{\Theta}_i$ is the value of Θ_i according to the fitted model.

(ii) If the ith data value O_i follows a Poisson distribution with mean and variance of E_i then standardized residuals are

$$R_i = (O_i - \hat{E}_i)/\sqrt{\hat{E}_i},$$

where $\hat{E}_i$ is the value of E_i according to the fitted model.

3.5 MULTIPLE TESTS AND CONFIDENCE INTERVALS

Often the researcher finds the need to perform several significance tests, or construct several confidence intervals at the same time using the same data. A problem then arises because error probabilities mount up. For example, if ten independent significance tests are carried out at the 5% level of significance, with null hypotheses being true, then the probability of one or more results being significant is $1 - 0.95^{10} = 0.40$. The researcher cannot therefore be sure how to react to one or two significant results out of the ten tests.

If the ten tests are not independent then the probability of getting at least one significant result will not be 0.40. Nevertheless, the principle applies that the more tests that are carried out, the more likely it is that there will be at least one significant result by chance when the null hypothesis is true for all the tests. Similarly, the probability of one or more confidence intervals not including population values increases with the number of these intervals that are constructed.

Perhaps the simplest way to overcome these problems involves making use of the Bonferroni inequality, which says that if n hypothesis tests are carried out simultaneously, each at the $100(\alpha/n)\%$ level, then the probability of declaring any result significant is α or less when the null hypotheses are all true. Likewise, the inequality says that if n confidence intervals are constructed, each with confidence level $100(1-\alpha/n)\%$, then the probability that all the intervals will include the true value of the population parameter is $1-\alpha$ or more. Thus if ten tests are carried out using the $(5/10)\% = 0.5\%$ level of significance, then the overall probability of declaring a result significant in error is 5% or less. Also, if confidence intervals for ten population parameters are constructed using a $(100 - 5/10)\% = 99.5\%$ confidence level for each one, then the probability that all the intervals will contain the true population parameter values will be 0.95 or more.

4 Studies with resources defined by several categories

In this chapter designs and data analysis procedures are reviewed for studies of resource selection where each resource unit is classified into one of several categories. These studies are the simplest that can be carried out, and are also the ones that are used most often. They are therefore worth considering in their own right, although they can be thought of as special cases of the more general types of study design that are discussed in later chapters.

4.1 INTRODUCTION

This chapter deals with the estimation of selection ratios as a means of studying selection in situations where each resource unit can be classified into one of several distinct categories. These selection ratios, which are equivalent to the forage ratios of Table 1.1, are defined such that for each resource unit the value of the ratio is proportional to the probability of that unit being utilized, given that the selecting organism has unrestricted access to the entire distribution of available units. The ratios can therefore be thought of as special cases of resource selection functions. Selection ratios are not the only way to analyse the results from selection experiments with resources in several categories, and a number of alternatives have been reviewed briefly in Chapter 1.

4.2 SAMPLING DESIGNS AND PROTOCOLS

Three general designs for recording data and three sampling protocols have been discussed in Chapter 1. These are summarized again here for convenience.

The type of study design depends on whether or not results are recorded for individual animals. With design I, data on the availability of resource units and their use are recorded at the population level. Hence any inferences can concern only the overall use of resources by all animals. With design II, data on the use of resources is recorded for each of several animals but availability is measured at the population level, and implicitly assumed to be approximately the same for each animal. With design III, data on both use and availability is collected for each of several animals. With designs II and III we assume that the animals used are selected independently with equal probability from a single population so that the animals provide the replication that is needed in order to make inferences concerning the use of resources by the population of animals.

With sampling protocol A (SPA), selection is studied by comparing a sample of used resource units with a sample or census of available resource units. With sampling protocol B (SPB), the comparison is between a sample of unused resource units and a sample or census of available units. With sampling protocol C (SPC) the comparison is between a sample of used resource units and a sample of unused resource units.

In this chapter we concentrate on the consideration of studies using SPA because these are by far the most common when resource units are considered to be divided into several categories on the basis of one factor such as habitat type. However, in passing, we note some important considerations with the other two types of sampling protocol.

First, if SPB is employed then the selection indices that can be estimated give the relative probabilities of different types of unit being unused, instead of relative probabilities of them being used. These indices may not be as satisfactory as indices for use, but using them it is still possible, for example, to order resource units in terms of their estimated probability of use.

When SPC is employed it may well be possible to argue that taking a sample or census of unused resource units is for all intents and purposes the same as taking a sample or census of available units because the proportion of available units that are actually used is so small. In that case, a study using SPC can be treated as one using SPA with a negligible error.

4.3 RATIOS OF RANDOM VARIABLES

Ratios of random variables arise in several places in this chapter. It is therefore useful to review some results concerning the estimation of ratios and differences between these ratios, since these results are not ordinarily available in introductory textbooks on statistics.

Let (Y, X) denote a pair of random variables measured on the units in a population. Suppose that there is interest in the estimation of $R = \mu_Y/\mu_X$, the ratio of the population mean of Y to the population mean of X, using data $(y_1,x_1),\ldots,(y_n,x_n)$ from a random sample of n units.

The estimator of R that is usually recommended is

$$\hat{R} = \bar{y}/\bar{x}, \tag{4.1}$$

where $\bar{x}$ and $\bar{y}$ are the sample means for X and Y. This is used in preference to

$$\hat{R}' = \sum_{j=1}^{n} (y_j/x_j)/n,$$

the sample mean of Y/X, since $\hat{R}$ will usually have less bias and a smaller variance that $\hat{R}'$.

The variance of $\hat{R}$ can be estimated by

$$\text{var}(\hat{R}) = \hat{R}^2\{(s_Y/\bar{y})^2 + (s_X/\bar{x})^2 - 2r_{XY}s_Ys_X/(\bar{y}\bar{x})\}/n, \qquad (4.2)$$

where $\bar{y}$ and s_Y are the sample mean and standard deviation for the Y values, $\bar{x}$ and s_X are the sample mean and standard deviation for the X values, and r_{XY} is the sample correlation coefficient for X and Y. Alternatively, it is often convenient to estimate the variance of $\hat{R}$ using the equation

$$\text{var}(\hat{R}) = \{\sum_{j=1}^{n}(y_j - \hat{R}x_j)^2/(n-1)\}\{1/(n\bar{x}^2)\}, \qquad (4.3)$$

(Cochran, 1977, p. 153).

In many applications of ratios there is no random sample of (x_i,y_i) pairs from n units. Rather, what is available are two summary statistics, X and Y, with known or estimated variances and covariance. In that case the ratio of the means can be estimated by $\hat{R} = y/x$ and its variance can be approximated by

$$\text{var}(\hat{R}) = \hat{R}^2[\text{var}(Y)/y^2 + \text{var}(X)/x^2 - 2\text{cov}(X, Y)/(yx)], \qquad (4.4)$$

where $\text{var}(X)$ and $\text{var}(Y)$ are the variances for X and Y, respectively, and $\text{cov}(X, Y)$ is the covariance between Y and X. Of course, if X and Y come from independent data then $\text{cov}(X, Y) = 0$.

Another situation that occurs in this chapter is where the difference between two ratios of the form of equation (4.1) is considered. Thus consider the random variable

$$\hat{D} = \hat{R}_1 - \hat{R}_2 = y_{1+}/x_{1+} - y_{2+}/x_{2+}, \qquad (4.5)$$

where $y_{1+} = y_{11} + y_{12} + \dots + y_{1n}$, $x_{1+} = x_{11} + x_{12} + \dots + x_{1n}$, $y_{2+} = y_{21} + y_{22} + \dots + y_{2n}$, and $x_{2+} = x_{21} + x_{22} + \dots + x_{2n}$, and where x_{1j}, y_{1j}, x_{2j} and y_{2j} are observations made on the jth unit in a random sample of n units. Here it can be shown using a standard Taylor series method (Manly, 1985, p. 408) that the variance of $\hat{D}$ can be approximated by

$$\text{var}(\hat{D}) = \{n/(n-1)\} \sum_{j=1}^{n} \{(y_{1j}-\hat{R}_1x_{1j}) / x_{1+} - (y_{2j}-\hat{R}_2x_{2j}) / x_{2+}\}^2. \qquad (4.6)$$

A further possibility here is that the difference being considered has the simpler form

$$\hat{D} = x_1/y_1 - x_2/y_2,$$

where X_1, Y_1, X_2 and Y_2 are random variables with known or estimated variances and covariances. The Taylor series approximation method then yields

$$\mathrm{var}(\hat{D}) = \{1/y_1^2\}\mathrm{var}(X_1) + \{x_1^2/y_1^4\}\mathrm{var}(Y_1) + \{1/y_2^2\}\mathrm{var}(X_2)$$
$$+ \{x_2^2/y_2^4\}\mathrm{var}(Y_2) - 2\{x_1/y_1^3\}\mathrm{cov}(X_1,Y_1)$$
$$- 2\{1/(y_1y_2)\}\mathrm{cov}(X_1,X_2) + 2\{x_2/(y_1y_2^2)\}\mathrm{cov}(X_1,Y_2)$$
$$+ 2\{x_1/(y_1^2 y_2)\}\mathrm{cov}(Y_1,X_2) - 2\{(x_1x_2)/(y_1^2 y_2^2)\}\mathrm{cov}(Y_1,Y_2)$$
$$- 2\{x_2/(y_2^3)\}\mathrm{cov}(X_2,Y_2). \tag{4.7}$$

This looks like a complicated equation. However, in practice some of the covariances may be zero, leading to some simplification.

With a small population it is desirable to apply a finite population correction to equations (4.2), (4.3) and (4.6). This involves multiplying the right-hand sides of these equations by the finite population correction $(1 - n/N)$, where N is the population size.

4.4 CHI-SQUARED TESTS

Use is made of chi-squared goodness of fit tests at various places in this chapter to test for significant selection, or to test for whether different animals are using resources differently. The form of test statistic that is most commonly used for this purpose is the Pearson statistic which takes the form

$$X_P^2 = \sum (O_i - E_i)^2 / E_i,$$

where O_i is an observed sample frequency, E_i is the expected value of O_i according to the hypothesis being considered, and the summation is over all the data frequencies. However, for most purposes we choose instead to use the log-likelihood statistic

$$X_L^2 = 2\sum O_i \log_e(O_i/E_i),$$

where again the summation is over all the data frequencies.

Both statistics have the same number of degrees of freedom and have chi-squared distributions for large samples if the null hypothesis being tested is correct. Furthermore, in practice X_P^2 and X_L^2 will give very similar values unless either the expected frequencies are very small or the differences between the observed and expected frequencies are very large. Therefore, it will usually not make much difference which statistic is used in terms of whether results are significant or not.

Our preference for using X_L^2 rather than X_P^2 is based on the fact that this is justified by the general theory of log-likelihood tests described in section 3.3, and this theory is used extensively in the chapters that follow. Thus all the chi-squared statistics that are used in the present chapter can be thought of as differences between maximized log-likelihoods for fitted models based on different assumptions.

4.5 DESIGN I WITH KNOWN PROPORTIONS OF AVAILABLE RESOURCE UNITS

The most common sampling plan used to study resource selection is the special case where (i) there is no unique identification of data collected from different animals; (ii) the proportions of available units in different resource categories are known; and (iii) a random sample of used resource units is taken. This is then design I, with sampling protocol A.

Suppose that this is the situation, and let the following notation apply: A_i = the number of available resource units in category i, for i = 1, 2,..., I; A_+ = the size of the total population of available resource units; $\pi_i = A_i/A_+$, the proportion of available resource units that are in category i; U_i = the number of used resource units in category i in the population; U_+ = the total number of used resource units in the population; u_i = the number of resource units in category i in the sample of used units; u_+ = the total number of used resource units sampled; $o_i = u_i/u_+$, the proportion of the sample of used resource units that are in category i; and w_i^* = the proportion of the population of available resource units in category i that are used (the resource selection probability function).

$$U_i = A_+ \pi_i w_i^*, \tag{4.8}$$

so that the selection probability for category i is

$$w_i^* = U_i/(A_+ \pi_i).$$

The proportion of used resource units in the population that are in category i is U_i/U_+, where this can be estimated by the proportion of resource units in this category in the sample of used units (o_i). Hence, an estimate of U_i is $o_i U_+$, and an estimate of the resource selection probability function is

$$\hat{w}_i^* = (o_i U_+)/(A_+ \pi_i).$$

In practice, it is likely that neither of the population totals U_+ nor A_+ will be known. However, the selection ratio

$$\hat{w}_i = o_i/\pi_i \tag{4.9}$$

can be calculated, where this is $\hat{w}_i^*$ multiplied by the unknown constant A_+/U_+.

Selection ratios are an old intuitive approach to analysis of resource selection related to a single categorical variable but generally have not been recognized as giving rise to relative probabilities of selection. The names used for o_i/π_i have varied from forage ratios (Hess and Swartz, 1940), and selectivity indices (Manly et al., 1972), to preference indices (Hobbs and Bowden, 1982).

A useful way of presenting selection ratios is with them standardized so that they add to 1. This leads to the standardized selection ratio

$$B_i = \hat{w}_i / (\sum_{i=1}^{I} \hat{w}_j), \tag{4.10}$$

which has the interpretation of being the estimated probability that a category i resource unit would be the next one selected if somehow it was possible to make each of the types of resource unit equally available. To understand this interpretation, suppose that a used resource unit is randomly chosen from the population of used units. Then, from equation (4.8), the probability of this unit being in resource category i is

$$U_i/U_+ = A_+\pi_i w_i^*/(\sum A_+\pi_j w_j^*)$$
$$= \pi_i w_i^*/(\sum \pi_j w_j^*)$$
$$= \pi_i w_i/(\sum \pi_j w_j),$$

where the last line follows because the value of the right-hand side of the equation is not affected by multiplying all of the w_i^* values by a constant. Hence, if all categories are equally available, so that $\pi_1 = \pi_2 = ... = \pi_I$, then

$$U_i/U_+ = w_i/(\sum w_j).$$

It follows that B_i of equation (4.10) gives the estimated probability that a randomly selected used resource unit will be in category i if all categories are equally frequent in the original population of available resource units (Manly *et al.*, 1972).

4.5.1 Example 4.1 Habitat selection by moose

The situation considered in example 2.1 is an example of the type of design just described. It may be recalled that Neu *et al.* (1974) considered selection of habitat by Moose (*Alces alces*) on a site surrounding the Little Sioux Burn in northeastern Minnesota during the winter of 1971-72, and collected the data shown in Table 2.1. The proportion of acreage in the ith habitat type, π_i, was determined by planimeter. Locations of moose and moose tracks were plotted on a map of the study area during aerial surveys. Each group of moose or moose tracks was considered as an independent observation in the sample of 'used' points.

   Table 4.1 shows the calculations for the selection ratios of equations (4.9) and (4.10) for this example. It can be seen that the edge habitat in the burn appears to have been selected with about three times the probability of the interior habitat in the burn ($B_2 = 0.326$, $B_1 = 0.110$). Similarly, the edge habitat out of the burn is about four times as likely to be selected as is the habitat further from the burn ($B_3 = 0.433$, $B_4 = 0.131$). However, edge habitat in the burn and edge habitat out of the burn are selected with approximately equal probability ($B_2 = 0.326$, $B_3 = 0.433$). The same relationships hold between the unstandardized $\hat{w}_i$ values. Procedures to test that there are no statistically significant differences between these pairs of values are discussed below.

Table 4.1 Estimation of selection indices for the occurrence of moose tracks on burned, unburned, and peripheral portions of a 33 200 acre area surrounding the Little Sioux Burn in northeastern Minnesota in the winter of 1971–72

Habitat	Population proportion (π_i)	Sample count (u_i)	Used sample proportion (o_i)	Selection index $(\hat{w}_i)$	Standardized index (B_i)
In burn, interior	0.340	25	0.214	0.629	0.110
In burn, edge	0.101	22	0.188	1.866	0.326
Out of burn, edge	0.104	30	0.256	2.473	0.433
Out of burn, further	0.455	40	0.342	0.750	0.131
Total	1.000	117	1.000	5.718	1.000

4.5.2 Example 4.2 Selection of escape cover by quail

For a second example, consider the selection ratios in Table 4.2; these are computed for a subset of data on selection of escape cover by California quail (*Callipepla californica*) from a study by Stinnett and Klebenow (1986). It can be seen that the shrubland habitat was estimated to be selected with about twice the probability of riparian habitat ($B_4 = 0.034$, $B_5 = 0.061$), and field border was approximately 30 times more likely to be selected than was riparian ($B_6 = 0.889$).

Table 4.2 Relative probabilities of selection of escape cover by quail

Escape cover	Sample count (u_i)	Expected count $(\pi_i u_+)$	Population proportion (π_i)	Selection ratio $(\hat{w}_i)$	Standardized ratio (B_i)
Pasture	0	23.5	0.362	0.000	0.000
Disturbed habitat	0	4.3	0.066	0.000	0.000
Farmstead	2	3.7	0.057	0.540	0.016
Riparian	19	16.2	0.249	1.174	0.034
Shrubland	36	17.0	0.262	2.114	0.061
Field border	8	0.3	0.004	30.769	0.889
Total	65	65.0	1.000	35.597	1.000

4.6 TESTS ON PROPORTIONS OF USED UNITS

The standard analysis in much of the recent literature for the situation where population proportions of resource categories are known and a sample of used units is taken has been to conduct a chi-squared test of the null hypothesis that animals are randomly selecting habitat in proportion to availability. If the chi-squared test is significant then it is followed by the computation of simultaneous confidence intervals for the population proportions of used resources of different types and the comparison of these with the available proportions (Neu *et al.*, 1974; Byers *et al.*, 1984).

The chi-squared statistic usually used takes the form

$$X_P^2 = \sum_{i=1}^{I}(u_i - u_+\pi_i)^2/(u_+\pi_i),$$

with I-1 degrees of freedom, where I is the number of resource categories. Thus the observed number of used resource units of type i (u_i) is compared with the expected number ($u_+\pi_i$) under the hypothesis of no selection. If X_P^2 is significantly large when compared to the percentage points of the chi-squared distribution then this indicates that there is a departure from the null hypothesis that selection is random.

As discussed in section 4.4, we choose to use the alternative statistic

$$X_L^2 = 2 \sum_{i=1}^{I} u_i \log_e \{u_i/(u_+\pi_i)\} \tag{4.11}$$

on the grounds that this is consistent with the use of log-likelihood tests in the chapters that follow. In practice X_P^2 and X_L^2 will give almost the same numerical results.

The condition for the chi-squared test to be valid is the usual one that the expected frequencies should be five or more. If this condition is not met then the test may still be valid, but obviously the outcome of the test should be treated with a certain amount of reservation.

The sample proportion o_i will have a standard error that is approximately given by

$$se(o_i) = \sqrt{\{o_i(1-o_i)/u_+\}}. \tag{4.12}$$

Hence approximate $100(1-\alpha)\%$ confidence intervals for the population proportions of used resource units of different types can be taken as

$$o_i \pm z_{\alpha/2}\sqrt{\{o_i(1-o_i)/u_+\}}, \tag{4.13}$$

where $z_{\alpha/2}$ is the percentage point of the standard normal distribution that is exceeded with probability $\alpha/2$. Since there are I types of resource unit, a Bonferroni adjustment to the confidence level is suggested, as discussed in section 3.5. In particular, if the value of α is set at 5%/I then there will be a probability of about 0.95 that all i confidence intervals will include their respective population ratios.

The confidence intervals defined by (4.13) will have the required level of significance providing that all of the sample proportions o_i are approximately normally distributed, and the standard errors calculated using equation (4.12) are close to the true standard errors. Since u_i should have a binomial distribution with mean estimated by u_+o_i and variance estimated by $u_+o_i(1-o_i)$, standard statistical theory suggests that these conditions will be met providing that $u_+o_i(1-o_i)$ is greater than about 5. This will occur if the observed number of used units exceeds five for each resource category. Basically, then, this is the same requirement as was mentioned earlier for the chi-squared test.

It may of course happen that there are fewer than five resource units in some categories. In that case the corresponding confidence intervals must be regarded as merely indicative of the level of sampling errors to be expected.

So far in this discussion on variances, tests of significance and confidence intervals it has been implicitly assumed that u_+, the sample size for the used resource units, is fixed in advance of sampling, so that u_i has a binomial distribution. However, in practice it will often be the case that researchers basically take whatever sample size they can get. This raises the question about the validity of inferences when sample sizes are random variables.

There are two facts here that justify the use of the equations given above. First, it can be argued that the inferences made are conditional on the observed sample size, and therefore rely on probability statements that apply for the restricted set of possible sets of data that can be obtained subject to the sample size observed. Second, it can be argued that if the total sample sizes is not fixed then the u_i values can be expected to have independent distributions that are approximately Poisson. It can then be shown that the statistic X_L^2 of equation (4.11) will still approximately have a chi-squared distribution with $I-1$ degrees of freedom, and equation (4.12) will still be a valid estimate of the standard error of o_i. Consequently, all the inference procedures will still be valid. Furthermore, a reasonable condition for the chi-squared and normal distribution approximations to hold will still be that the number of used units sampled is five or more in each resource category.

4.6.1 Example 4.2 (continued) Selection of escape cover by quail

Consider again the data in Table 4.2 on the escape cover used by quail. Here the X_L^2 statistic of equation (4.11) is 112.35, with five degrees of freedom, which is highly significant when compared with the percentage points of the chi-squared distribution. Since three of the expected frequencies are less than five, the interpretation of this result requires some caution. In the analysis below, we consider the consequences of dropping these rarely used habitats from the data set. At any rate, there does seem to be very clear evidence of selection because X_L^2 is very large.

If $\alpha = 0.05/6 = 0.0083$ is used with the six confidence intervals (4.13) then it is possible to be approximately 95% confident that all the limits contain the population values. The z_α value is then 2.64, and the limits are as shown in Table 4.3. It can be seen that shrubland and field border were apparently selected significantly more often than is expected from the population proportions of these habitats (e.g., for shrubland, $\pi_5 = 0.262$ is below the lower limit of the confidence interval 0.391 to 0.717). However, the fact that three observed frequencies are less than five must lead us to treat the significance of this result with some reservations.

The reason for having this example with low observed frequencies is that it provides a good illustration of one problem that can occur when the proportions of used resource units in different categories are compared with the proportions available. Thus suppose that the researchers consider dropping pasture and disturbed habitat from the study because these types are known to be rarely selected. As shown in Table 4.3, the observed proportion of times that the riparian habitat is chosen (0.292) exceeds the expected proportion (0.249) when six habitat types are considered. Thus this habitat seems to be favoured to some extent. However, as shown in Table 4.4, if pasture and disturbed habitats are dropped from the analysis then the observed proportion of riparian choices stays the same at 0.292 but the expected proportion increases to 0.435. Now the riparian habitat seems to be avoided.

Table 4.3 Bonferroni confidence intervals for population proportions of resource units used by quail

Escape cover	Population proportion (π_i)	Used sample proportion (o_i)	Bonferroni confidence limits Lower	Upper
Pasture	0.362	0.000	–	–
Disturbed habitat	0.066	0.000	–	–
Farmstead	0.057	0.031	0.000[*]	0.088
Riparian	0.249	0.292	0.143	0.441
Shrubland	0.262	0.554	0.391	0.717
Field border	0.004	0.123	0.015	0.231

[*]For farmstead a negative lower limit has been changed to 0.000. The limits for this habitat are unreliable because of the low sample count of used resource units.

This effect of a resource category switching from preferred to avoided, or vice-versa, is not uncommon when a large but seldom used resource category is removed from the analysis. It means, of course, that the interpretation of data may depend rather crucially on decisions that are made concerning what types of habitat are 'available' to an animal.

One of the advantages of working with selection ratios is that this effect is largely avoided. Thus the selection ratios for farmstead, riparian, shrubland and field border that are shown in Table 4.2 (with pasture and disturbed habitats included) are proportional to the selection ratios that are shown in Table 4.5 (with pasture and disturbed habitats excluded). This is seen by the fact that the standardized ratios B_i are identical in Tables 4.2 and 4.4.

Table 4.4   Selection of escape cover by quail with the pasture and disturbed habitat types removed

Escape cover	u_i	π_i	o_i	$\hat{w}_i$	B_i
Farmstead	2	0.100	0.031	0.308	0.016
Riparian	19	0.435	0.292	0.677	0.034
Shrubland	36	0.458	0.554	1.209	0.061
Field border	8	0.007	0.123	17.582	0.889
Total	65	1.000	1.000	19.766	1.000

4.7 INFERENCES CONCERNING SELECTION RATIOS

The standard error of $\hat{w}_i$ can be approximated by

$$se(\hat{w}_i) = se(o_i/\pi_i) = \sqrt{\{o_i(1-o_i)/(u_+\pi_i^2)\}}. \tag{4.14}$$

Hence an approximate $100(1-\alpha)\%$ confidence interval for a single selection ratio w_i is of the form

$$\hat{w}_i \pm z_{\alpha/2}se(\hat{w}_i), \tag{4.15}$$

where $z_{\alpha/2}$ is the critical value of the standard normal distribution corresponding to an upper tail area of $\alpha/2$.

The condition for the limits (4.15) to have about the right level of confidence is the same as the condition for the binomial proportion o_i to be approximately normally distributed, since if this is the case then o_i/π_i will also be approximately normally distributed. As discussed earlier, a reasonable requirement is therefore that the number of used resource units is five or more in all resource categories. If this condition does not hold for some categories then the intervals are somewhat suspect for these categories, but will be reliable for the others.

The selection coefficient $\hat{w}_i$ is significantly different from 1 if the confidence interval for w_i does not contain the value 1. Alternatively, an approximate test for the significance of $\hat{w}_i$ involves comparing $(\hat{w}_i - 1)/se(\hat{w}_i)$ with critical values for the standard normal distribution, or $\{(\hat{w}_i-1)/se(\hat{w}_i)\}^2$ with critical values for the chi-squared distribution with one degree of freedom.

There is the possibility for confusion here because $se(\hat{w}_i)$ can be estimated either on the assumption that selection may occur, or on the assumption that there is no selection. In the first case, equation (4.14) is the appropriate one to use, with the proportion of category i resource units in the population of used units being approximated by the sample proportion o_i. In the second case, the expected value of o_i becomes equal to π_i, and equation (4.14) can be replaced by

$$se(\hat{w}_i) = \sqrt{\{(1-\pi_i)/(u_+\pi_i)\}}. \tag{4.16}$$

It can be argued that this last equation is the one to use when testing the hypothesis that there is no selection since it gives the standard error exactly if this hypothesis is true.

In practice, the researcher is usually interested in the entire set of I selection ratios, w_i, for i from 1 to I. When this is the case, approximate simultaneous confidence intervals or tests can be constructed by use of the Bonferroni inequality that has been discussed in sections 3.5 and 4.6. Thus (subject to the conditions already mentioned for the normal approximation to be valid), it is possible to be $100(1-\alpha)\%$ confident that the intervals for all I selection ratios contain their respective true ratios if $z_{\alpha/2}$ is replaced in (4.15) by $z_{\alpha/(2I)}$, the standard normal table value corresponding to an upper tail probability of $\alpha/(2I)$. Similarly, if the significance levels used for testing $(\hat{w}_i-1)/se(\hat{w}_i)$ against the standard normal distribution is α/I then there will be a probability of only approximately α of getting any result significant when there is no selection.

The discussion at the end of section 4.6 concerning the validity of inferences when sample sizes are not fixed in advance carries over to inferences on selection ratios. Hence all the results that have been given in the present section apply equally well if u_+, the total sample size of used units, is not fixed in advance.

4.7.1 Example 4.2 (continued) Selection of escape cover by quail

Consider again Stinnett and Klebenow's (1986) study of the escape cover selected by quail with the data shown in Table 4.2. Since there are I = 6 habitat types, it is appropriate to test each selection ratio for significance using the $(5/6)\% = 0.8\%$ level of significance. For example, the selection ratio for shrubland as escape cover is $\hat{w}_5 = 2.114$. Assuming that there is no selection, the standard error associated with this ratio is given by equation (4.16) to be $se(\hat{w}_5) = \sqrt{\{(1-0.262)/(65\times0.262)\}} = 0.208$. Hence the chi-squared statistic with one degree of freedom to test for selection is

$$(\hat{w}_5-1)^2/se(\hat{w}_5)^2 = (2.114 - 1)^2/0.208^2 = 28.64.$$

Since this is significantly large at the 0.8% level, this indicates that shrubland is used more than is expected from the availability of this habitat. Similar calculations suggest that there is selection against pasture, and selection for shrubland and field border.

4.8 COMPARISON OF SELECTION RATIOS

Assuming that the sample of used units is random, the observed counts for resource categories u_1 to u_I will follow a multinomial distribution, conditional on the total sample size u_+ being regarded as being fixed in advance. This means that the sample proportions o_1 to o_I will be multinomial proportions with estimated variances $var(o_i) = o_i(1-o_i)/u_+$ and estimated covariances $cov(o_i,o_j) = -o_io_j/u_+$. The variance of the difference between two selection ratios can therefore be estimated by

$$\text{var}(\hat{w}_i - \hat{w}_j) = \text{var}(o_i/\pi_i) - 2\text{cov}(o_i/\pi_i, o_j/\pi_j) + \text{var}(o_j/\pi_j),$$

$$= o_i(1-o_i)/(u_+\pi_i^2) - 2o_i o_j/(u_+\pi_i\pi_j)$$

$$+ o_j(1-o_j)/(u_+\pi_j^2). \tag{4.17}$$

Using this equation, the null hypothesis of no difference in the probabilities of selection of the ith and the jth categories, i.e., that $w_i = w_j$ for $i \neq j$, can be tested by comparing the statistic $(\hat{w}_i - \hat{w}_j)^2/\text{var}(\hat{w}_i - \hat{w}_j)$ with the critical values of the chi-squared distribution with one degree of freedom. Also, an approximate $100(1-\alpha)\%$ confidence interval for the difference $w_i - w_j$ is given by

$$(\hat{w}_i - \hat{w}_j) \pm z_{\alpha/2}\text{se}(\hat{w}_i - \hat{w}_j). \tag{4.18}$$

The validity of these confidence intervals depends on the assumption that the estimators $\hat{w}_i$ are normally distributed, which in turn depends on the sample proportions o_i being normally distributed. As discussed before, a reasonable requirement for this to hold is that the number of used units should be five or more for each resource category. Hence if this condition is not met for some of the resource categories then any of the confidence intervals (4.18) involving these categories must obviously be treated with some reservations.

Using Bonferroni's inequality, a procedure very similar to that used to compare means in analysis of variance can be suggested for comparing the selection ratios. This involves ranking the selection ratios from the smallest to the largest and comparing them two at a time by use of the confidence intervals (4.18), replacing $z_{\alpha/2}$ by $z_{\alpha/(2I')}$ where I' is the total number of comparisons being made (the number of combinations of I categories taken two at a time). The selection ratios w_i and w_j are then declared significantly different if the confidence interval for $w_i - w_j$ does not contain zero (subject to the requirement that u_i and u_j should both be five or more).

As was the case with inferences concerning sample proportions of used resource units of different types (section 4.6) and inferences concerning individual selection ratios (section 4.7), all the results in the present section can be justified for use in situations where u_+, the total number of used units sampled, is not fixed in advance. In particular, the variance equation (4.17) can be shown to give a valid estimate of variance if the u_i values follow independent Poisson distributions.

4.8.1 Example 4.2 (continued) Selection of escape cover by quail

The results of the tests just described are shown in Table 4.5 for the selection ratios obtained from Stinnett and Klebenow's (1986) study of selection of escape cover by quail. The differences between selection ratios that are connected with a vertical bar have an approximate $(100 - 5/15)\% = 99.67\%$ confidence interval that includes zero. Since there are 15 pairwise comparisons, this procedure gives about a 0.95 probability that all the confidence limits include their respective population differences between selection ratios.

From Table 4.5 it can be seen that:

(a) the probabilities of selection for shrubland and field border are significantly different, but both are significantly larger than the probability of selection for the other habitat types;
(b) the probability of selection of riparian is not significantly different from the probability of selection of farmland; and
(c) there are no significant differences between the probabilities of selection for pasture, disturbed habitats and farmstead.

Since the sample counts for pasture, disturbed habitats and farmstead are less than five, the comparisons involving these three categories must be regarded as indicative only.

Table 4.5 Comparison between selection ratios using confidence limits chosen so that the probability of all the pairwise intervals including the population difference is 0.95. Comparisons involving pasture, disturbed habitat and farmstead are unreliable because of low or zero counts of used units.

Escape cover	w_i	
Pasture	0.000	*
Disturbed habitat	0.000	
Farmstead	0.540	
Riparian	1.174	
Shrubland	2.114	I
Field border	30.769	I

*Selection ratios connected by a vertical bar are not significantly different at the 0.95 probability level.

4.9 DESIGN I WITH ESTIMATED PROPORTIONS OF AVAILABLE RESOURCE UNITS

Often studies using design I and sampling protocol A involve estimating the proportions π_i of available units using a random sample of available resource units. For example, Marcum and Loftsgaarden (1980) estimated the proportion of different types of habitat in a study area by locating a number of random points on a map of the study area and counting the number of points 'hitting' each type.

In studies of this type, the usual two way chi-squared test can be used to see whether the sample proportion of used units is significantly different from the sample proportion of available units. Thus, assume that a random sample of u_+ used units yields u_i units in category i, and a random sample of m_+ available units yields m_i in the same category, for i from 1 to I. Then the usual chi-squared statistic can be written as

$$X_P^2 = \sum_{i=1}^{I} [\{u_i - E(u_i)\}^2/E(u_i) + \{m_i - E(m_i)\}^2/E(m_i)],$$

where $E(u_i) = (m_i+u_i)u_+/(u_++m_+)$ is the expected value of u_i, and $E(m_i) = (m_i+u_i)m_+/(u_++m_+)$ is the expected value of m_i, on the hypothesis of no selection. If this statistic is significantly large when compared with the chi-squared distribution with I–1 degrees of freedom then there is evidence that selection is occurring. The usual conditions for the validity of the chi-squared test apply, so that it should be reliable if all the expected frequencies are five or more.

For the reasons discussed in section 4.4, we choose to replace X_P^2 with the log-likelihood statistic

$$X_L^2 = 2 \sum_{i=1}^{I} \{u_i\log_e\{u_i/E(u_i)\} + m_i\log_e\{m_i/E(m_i)\}\}, \tag{4.19}$$

which has the same degrees of freedom and validity conditions as X_P^2.

The sample of available resource units gives the estimator

$$\hat{\pi}_i = m_i/m_+, \tag{4.20}$$

of π_i, with estimated standard error

$$se(\hat{\pi}_i) = \hat{\pi}_i(1-\hat{\pi}_i)/m_+. \tag{4.21}$$

An estimator of w_i is therefore

$$\hat{w}_i = o_i/\hat{\pi}_i, \tag{4.22}$$

where, as before, $o_i = u_i/u_+$ is the proportion of category i resource units in the random sample of u_+ used units. The standard error of the estimated selection ratio $\hat{w}_i$ can be approximated from the general formula (4.4) for a ratio, which provides

$$\begin{aligned} se(\hat{w}_i) &= \hat{w}_i\sqrt{\{(1-o_i)/(o_iu_+) + (1-\hat{\pi}_i)/(\hat{\pi}_i m_+)\}}, \\ &= \hat{w}_i\sqrt{\{1/u_i - 1/u_+ + 1/m_i - 1/m_+\}}. \end{aligned} \tag{4.23}$$

There will zero covariance between o_i and $\hat{\pi}_i$ if they are estimated from independent samples.

Equation (4.23) can also be shown to be a valid estimate of the standard error of the estimated selection ratio for cases where the counts of used resource units u_1 to u_I and the counts of available resource units m_1 to m_I have independent Poisson distributions. Therefore, situations are covered where the total sample sizes u_+ and m_+ of used and available resource unit are not fixed in advance. The situation in this respect is therefore the same as for cases where population proportions of available resource units are known accurately.

Approximate simultaneous confidence intervals on the selection ratios can be constructed following the procedures suggested before. Thus the confidence intervals

$$\hat{w}_i \pm z_{\alpha/(2I)}se(\hat{w}_i) \tag{4.24}$$

can be considered, for i from 1 to I, where $z_{\alpha/(2I)}$ is the upper $100\alpha/(2I)$ percentage point for the standard normal distribution. The selection coefficient $\hat{w}_i$ is then declared significantly different from 1 if the confidence interval on w_i does not contain the value 1. The Bonferroni inequality suggests that this procedure will give a probability of approximately $1-\alpha$ that the population selection ratios will all be within their respective intervals, and that the probability of declaring any result significant will be approximately α if there is no selection.

The validity of the confidence intervals (4.24) depends on the standard errors from equation (4.23) being accurate, and the normality of the estimators $\hat{w}_i$. A minimum condition would seem to be that there are at least five resource units in each category both in the sample of used units and in the sample of available units, since this will at least ensure that the o_i and $\hat{\pi}_i$ values are approximately normally distributed, with reasonable estimates of their standard errors. However, the fact that $\hat{w}_i$ is a ratio of random variables suggests that some more stringent conditions may be needed. This matter is discussed further in the example that follows.

Estimation of the availabilities π_i by sampling protocols such as quadrat or transect sampling which do not use random points as the basic sampling unit will require a different equation for the estimation of the standard error of $\hat{w}_i$. Discussion of all possible cases is not realistic in this chapter. However, if an estimate of $se(\hat{\pi}_i)$ is available from the implemented sampling design then it can be substituted into equation (4.4) to obtain

$$se(\hat{w}_i) = \hat{w}_i \sqrt{\{(1 - o_i)/(o_i u_+) + se(\hat{\pi}_i)^2/\hat{\pi}_i^2\}}. \tag{4.25}$$

This equation can then be used to obtain confidence intervals for population selection ratios in the usual way.

The standard error of the difference between two selection ratios can be calculated using the general equation (4.7) by setting $x_1 = o_i$, $y_1 = \hat{\pi}_i$, $x_2 = o_j$ and $y_2 = \hat{\pi}_j$. Then assuming that o_i, o_j, $\hat{\pi}_i$ and $\hat{\pi}_j$ are simple proportions, with the first two of these quantities being independent of the second two, there are the following estimates of variances and covariances: $var(o_i) = o_i(1-o_i)/u_+$, $var(\hat{\pi}_i) = \hat{\pi}_i(1-\hat{\pi}_i)/m_+$, $cov(o_i,o_j) = - o_i o_j/u_+$ and $cov(\hat{\pi}_i,\hat{\pi}_j) = -\hat{\pi}_i\hat{\pi}_j/m_+$. In addition, covariances between o_i and $\hat{\pi}_j$ terms are zero, for all i and j. Using these results equation (4.7) produces the estimated variance

$$var(\hat{w}_i - \hat{w}_j) = \{ \hat{w}_i/\hat{\pi}_i + \hat{w}_j/\hat{\pi}_j - (\hat{w}_i - \hat{w}_j)^2 \}/u_+ \\ + \{ \hat{w}_i^2/\hat{\pi}_i + \hat{w}_j^2/\hat{\pi}_j - (\hat{w}_i - \hat{w}_j)^2 \}/m_+.$$

Bonferroni confidence intervals for the whole set of possible differences between selection ratios can be calculated as

$$(\hat{w}_i - \hat{w}_j) \pm z_{\alpha/(2I')} se(\hat{w}_i - \hat{w}_j),$$

where I' is the number of such differences. The procedure is exactly the same as described in section 4.8.

4.9.1 Example 4.3 Selection of forest canopy cover by elk

Marcum and Loftsgaarden (1980) used data from Marcum (1975) as an example where 200 random points were located on a map of the study area which contained a complex mosaic of forest-canopy cover, and compared with 325 elk (*Cervus elaphus*) locations in the same region. Four categories of habitat were considered, corresponding to canopy covers of 0%, 1–25%, 26–75% and 75–100%. The data are shown in Table 4.6, together with the estimated selection ratios, standard errors and confidence intervals that are obtained using equations (4.22) to (4.24). From equation (4.19), $X_L^2 = 21.9$, which is very highly significant when compared with percentage points of the chi-squared distribution with three degrees of freedom. There is therefore clear evidence of selection.

Table 4.6 Estimated percentages of different habitat types, and selection indices for forest canopy coverage classes available to elk[*]

Canopy cover class								Confidence limits	
	m_i	$\hat{\pi}_i$	u_i	$\hat{o}_i$	$\hat{w}_i$	B_i	$se(\hat{w}_i)$	Lower	Upper
0%	15	0.075[*]	3	0.009	0.120	0.038	0.077	0.000[†]	0.296
1–25%	61	0.305	90	0.277	0.908	0.289	0.127	0.624	1.191
26–75%	84	0.420	181	0.557	1.326	0.422	0.128	1.039	1.613
>75%	40	0.200	51	0.157	0.785	0.250	0.150	0.449	1.121
Total	200	1.000	325	1.000	3.139	1.000			

[*]The percentages available were estimated by a random sample of m = 200 points on a map of the study area. A sample of u_+ = 325 points selected by elk was used to estimate the proportions of elk observations in each class. The confidence limits for each selection ratio have a confidence level (100–10/4)% = 97.5% in order that there is a 0.9 probability that all four limits include the population selection ratio.
[†]A negative lower limit for the confidence interval for 0% has been replaced by 0.000 since negative values for the selection indices are impossible.

From the results in the table it appears that elk used the 0% canopy cover class significantly less than in proportion to its availability because the upper limit of the confidence interval for w_1 is below the value 1. Similarly, elk used the 26–75% canopy cover class significantly more than in proportion to availability because the lower limit of the confidence interval for w_3 is above one. There appears to be no significant selection either for or against the remaining two classes. These conclusions agree with the analysis of Marcum and Loftsgaarden (1980) based on simultaneous confidence intervals on the differences (π_i-o_i) rather than on the ratios $w_i = o_i/\pi_i$.

Since there were only three used points in the 0% canopy cover class, there is some question about the validity of the confidence interval for the selection ratio in this class. A small simulation experiment was therefore carried out to investigate this matter. What was done was to generate 500 sets of data similar to the observed data, with the expected sample frequencies of available and used resource units set equal to the observed frequencies, and the actual counts following independent Poisson distributions. Selection ratios and their standard errors were then estimated for each set of artificial data, together with the z-scores $z_i = (\hat{w}_i - w_i)/se(\hat{w}_i)$, where the 'true' selection ratios w_i were the values obtained from the original data.

The idea behind calculating the z-scores was that if these scores have a standard normal distribution then the confidence intervals (4.24) are valid. In fact, it was found that the z-scores had distributions that were very close to normal for the estimates of all four selection ratios, although there was a slight excess of values less than −3, counterbalanced by some lack of values of +3 or more. This is shown in Figure 4.1, which compares the distribution obtained for all the observed z-scores with the standard normal distribution.

The only real problem that was thrown up by the simulation was an occasional zero value of u_1. In these cases $\hat{w}_1 = 0$, $se(\hat{w}_1) = 0$, and z_1 becomes undefined. This occurred 12 times for the 500 sets of simulated data.

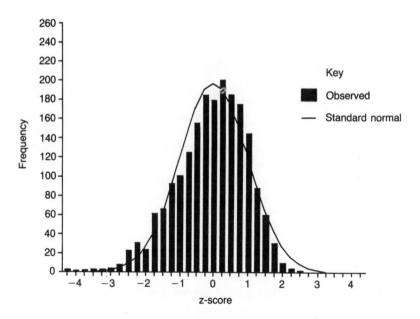

Figure 4.1 Comparison of the distribution of values of $z_i = (\hat{w}_i - w_i)/se(\hat{w}_i)$ with the standard normal distribution. If z_i follows a standard normal distribution then confidence limits (4.24) for population selection ratios will be reliable. The z_i values were obtained by simulating 500 sets of data similar to Marcum and Loftsgaarden's data on the selection of forest overstory cover by elk. Since there are four selective values, the distribution shown is for 2000 z-scores, less 12 cases where $u_1 = 0$ led to undefined values for z_1.

Overall, it seems that the confidence limits (4.24) are reasonably reliable even though there is a sample count of less than five used units in the 0% canopy cover class.

4.10 DESIGN II WITH SAMPLING PROTOCOL A

With sample design II, data are available on the selection of resource units by individual animals. For example, the use of resource categories might be estimated in the home ranges of individual animals. Thus, assume a sample of size n is obtained from the population of animals, the resource units used by the jth animal are sampled to estimate the proportions of different types of resource categories that it uses, and the universe of available resource units is sampled or censused to estimate the proportions of units in each of several categories.

With studies of this type we prefer to consider the animal as the primary sampling unit, and to base statistical inferences on the use of animals as replicates. Hence sampling the used units for the jth animal is viewed as 'subsampling' the primary sampling unit. We assume at all times that samples of animals or resource units are random samples.

4.11 CENSUS OF AVAILABLE RESOURCE UNITS

Suppose that the proportions of available resources in categories 1 to I are known to be π_1 to π_I for the entire study area, and that there is the following notation: u_{ij} = the number of type i resource units used by animal j; u_{i+} = the number of type i resource units used by all animals; u_{+j} = the total number of units used by animal j; and u_{++} = the total number of units used by all animals. In this situation there are two chi-squared tests that provide useful information about selection. First, a test can be carried out to see whether animals seem to be using resource categories in a similar way, irrespective of whether this is selective or not. This is the usual test for independence of the row and column categories in a two-way table, for which the log-likelihood test statistic is

$$X_{L1}^2 = 2 \sum_{j=1}^{n} \sum_{i=1}^{I} u_{ij} \log_e \{ u_{ij}/E(u_{ij}) \}, \tag{4.26}$$

where $E(u_{ij}) = u_{i+}u_{+j}/u_{++}$ is the expected number of units of type i used by the jth animal if that animal uses the resources in the same way as the other animals. If X_{L1}^2 is significantly large in comparison with the chi-squared distribution with $(I-1)(n-1)$ degrees of freedom then there is evidence that the animals are using resources differently, which implies that at least some of them are selective. As usual with a chi-squared test, it is desirable that all the expected frequencies should be five or more.

The second chi-squared test that can be considered is an overall test for selection which involves comparing the observed frequencies of resource categories used by different animals with expected frequencies calculated from the resources available. The log-likelihood test statistic is then

$$X_{L2}^2 = 2 \sum_{j=1}^{n} \sum_{i=1}^{I} u_{ij} \log_e \{u_{ij}/E(u_{ij})\},$$ (4.27)

where $E(u_{ij}) = \pi_i u_{+j}$ is the expected number of resource type i units used by the jth animal if use is proportional to availability. If this statistic is significantly large in comparison with the chi-squared distribution with $n(I-1)$ degrees of freedom then there is evidence of selection by at least some of the animals.

Because of the way that they are calculated, X_{L1}^2 must be less than or equal to X_{L2}^2, and the difference $X_{L1}^2 - X_{L2}^2$, with $I-1$ degrees of freedom, is a measure of the extent to which animals are on average using resources in proportion to availability, irrespective of whether they are selecting the same or not.

The selection ratio for the jth animal and the ith type of resource can be estimated by

$$\hat{w}_{ij} = u_{ij}/(\pi_i u_{+j}),$$ (4.28)

and there are two plausible estimators of the selection ratio w_i for the whole population of animals. First, the ratios for the n sampled animals can be averaged, to give

$$\hat{w}_i' = \sum_{j=1}^{n} \hat{w}_{ij}/n.$$

Second, totals for all animals can be used to give

$$\hat{w}_i = u_{i+}/(\pi_i u_{++}).$$ (4.29)

We recommend the use of $\hat{w}_i$ rather than $\hat{w}_i'$. This coincides with the usual practice for ratio estimation because ratios of means or totals generally have less bias and variance than means of ratios (section 4.3). Conditional on known values for π_i, the variance of $\hat{w}_i$ is estimated by a special case of equation (4.3) with $y_j = u_{ij}/\pi_i$ and $x_j = u_{+j}$. That is,

$$\text{var}(\hat{w}_i) = \{\sum_{j=1}^{n} (u_{ij}/\pi_i - \hat{w}_i u_{+j})^2/(n-1)\} \{n/u_{++}^2\}.$$ (4.30)

The estimates of the selection ratios are computed by pooling observations across all animals in the sample. However, the equation (4.30) takes variation in resource selection from animal to animal into account since the expression $\Sigma(u_{ij}/\pi_i - \hat{w}_i u_{+j})^2/(n-1)$ is an estimate of the variance of $u_{ij}/\pi_i - w_i u_{+j}$ in the population of animals.

Simultaneous Bonferroni confidence intervals for population selection ratios can be constructed with an overall confidence level of approximately $100(1-\alpha)\%$, so that the probability of all the intervals containing the true value is approximately $1-\alpha$. These intervals are of the form

$$\hat{w}_i \pm z_{\alpha/(2I)}se(\hat{w}_i), \tag{4.31}$$

where $z_{\alpha/(2I)}$ is the percentage point of the standard normal distribution corresponding to an upper tail probability of $\alpha/(2I)$, and I is the number of habitat types.

The validity of these confidence intervals depends on the standard error of $\hat{w}_i$ being well estimated, and $\hat{w}_i$ being normally distributed. Because of the complex nature of the estimation it is difficult to know precisely when this will occur. However, this question is addressed further in the example that follows.

The difference w_i-w_k between the selection ratios for resource units in categories i and k can be estimated by

$$\hat{w}_i-\hat{w}_k = (u_{i+}/\pi_i-u_{k+}/\pi_k)/u_{++}. \tag{4.32}$$

This is a ratio estimator of the general form discussed in section 4.3, and the variance can be estimated using equation (4.3) by setting $y_j = u_{ij}/\pi_i-u_{kj}/\pi_k$ and $x_j = u_{+j}$. This gives the result

$$var(\hat{w}_i-\hat{w}_k) = [\{n/(n-1)\}/u_{++}^2] \sum_{j=1}^{n} (u_{ij}/\pi_i-u_{kj}/\pi_k-\hat{w}_i\, u_{+j}+\hat{w}_k u_{+k})^2. \tag{4.33}$$

Again, confidence intervals for population differences can be constructed based on the Bonferroni inequality. These take the form

$$\hat{w}_i-\hat{w}_k \pm z_{\alpha/(2I')}se(\hat{w}_i-\hat{w}_k), \tag{4.34}$$

where $z_{\alpha/(2I')}$ is the value for the standard normal distribution that is exceeded with probability $\alpha/(2I')$, and I' is the number of differences that can be constructed between selection ratios. In this way, the probability that all the intervals will include their respective population ratios will be approximately $1-\alpha$.

4.11.1 Example 4.4 Habitat selection by bighorn sheep

Arnett *et al.* (1989) studied the selection of habitat types using a sample of six radio-tagged bighorn sheep (*Ovis canadensis*) in the Encampment River drainage of southeast Wyoming, with the proportions of ten habitat types available in the study area being measured from maps. A subset of their data covering the period August to the end of December 1988 is shown in Table 4.7. Each animal was 'subsampled' during this time to obtain a sample of habitat points used.

The test statistic of equation (4.26) is $X_{LI}^2 = 99.20$, with 40 degrees of freedom if the zero counts of riparian are ignored, or 45 degrees of freedom if these are included in the calculation with expected frequencies of zero. With either choice of degrees of freedom the statistic is very significantly large, indicating that the sheep were not using resources in the same way.

Table 4.7 Habitat type, proportion of study area in each type, and number of occasions a given bighorn sheep was observed in each type

Habitat	Available proportion	Use of habitats by sheep number 1	2	3	4	5	6	Total
Riparian	0.060	0	0	0	0	0	0	0
Conifer	0.130	0	2	1	1	0	2	6
Mt. shrub I	0.160	0	1	2	3	2	1	9
Aspen	0.150	2	2	1	7	2	4	18
Rock outcrop	0.060	0	2	0	5	5	2	14
Sage/bitterbrush	0.170	16	5	14	3	18	7	63
Windblown ridges	0.120	5	10	9	6	10	6	46
Mt. shrub II	0.040	14	10	8	9	6	15	62
Prescribed burns	0.090	28	35	40	31	25	19	178
Clearcut	0.020	8	9	4	9	0	19	49
	Total	73	76	79	74	68	75	445

The test statistic of equation (4.27) was much larger at $X_{L2}^2 = 785.54$, with 54 degrees of freedom, indicating very strong evidence indeed that the sheep were selective in their use of habitat. The difference $X_{L2}^2 - X_{L1}^2 = 686.34$, with 9 degrees of freedom (counting riparian frequencies) is an indication of the average level of selection, which is obviously very strong.

The interpretation of these test results must be tempered by the fact that many of the observed frequencies are less than five. However, the extreme levels of significance for the chi-squared statistics leave little room to doubt that selection took place, and varied from animal to animal. As in example 4.2, these data are presented to illustrate common problems which arise in real studies, that the procedures recommended are reasonably robust with respect to the low observed frequencies, and that the selection ratios computed in this chapter are not sensitive to ad hoc decisions concerning whether or not the rarely used habitat types (e.g., riparian and conifer) are included in the analysis. To investigate the robustness of the selection ratios and confidence limits to the low frequencies, a simulation experiment was conducted to assess the extent to which the confidence intervals (4.31) can be relied upon.

What was done was to simulate 215 sets of data with ten habitats and six sheep in such a way that the number of habitat i resource units used by animal j was a Poisson random variable with a mean value given by the observed frequency in Table 4.7. For each set of data, selection ratios and their estimated standard errors were calculated, and hence the z-scores $z_i = (\hat{w}_i - w_i)/se(\hat{w}_i)$, for i from 1 to 10. The 'true' selection ratios w_i were set equal to the estimates obtained from the data in Table 4.7 on the grounds that this was the 'population' for the simulations.

The riparian habitat was not chosen by any of the bighorn sheep for data shown in Table 4.7. Consequently, this habitat was not chosen in the simulations either. As a result, $w_1 = \hat{w}_1 = se(\hat{w}_1) = 0$, and z_1 is always undefined. The following comments therefore relate to the other z-scores only. The reason for expressing the results in terms of z-scores is that if the z_i values have approximately standard normal distributions then the confidence intervals (4.31) will be valid.

Although only 215 sets of data were simulated, this produced 1935 z-scores since there were nine values for each set of data. The nine values for one set of data are not, of course, independent. Nevertheless, a very clear pattern emerged suggesting that most z-scores are less variable than can be expected from the standard normal distribution, but there are occasional very extreme values that should not occur with the standard normal distribution. This is illustrated by Figure 4.2, which shows how the observed distribution for all 1935 z-scores compares with the standard normal distribution.

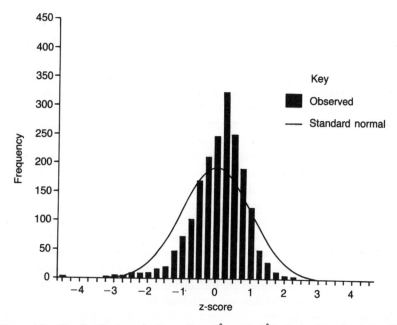

Figure 4.2   The distribution of values of $z_i = (\hat{w}_i - w_i)/se(\hat{w}_i)$ obtained by simulating 215 sets of data similar to the observed data that are shown in Table 4.7. It is required that z_i has a standard normal distribution for the confidence limits (4.31) to be valid.

This simulation study indicates that, on the whole, the confidence limits (4.31) are reliable, and, if anything, tend to be conservative. Furthermore, the occasional very extreme z-scores were associated with situations where only one animal used a resource category once. Therefore, these situations can readily be identified. Further simulations indicate that this statement is also true for the confidence intervals (4.34) for differences between selection ratios.

Having established the validity of our method for determining confidence intervals, these intervals can now be considered for the 'real' data. Table 4.8 contains the selection indices for the ten habitat types, together with Bonferroni confidence intervals computed from (4.31) using $\alpha = 0.1$, so that there is a probability of about 0.9 that all $I = 10$ intervals will contain their respective population selection ratios.

Table 4.8 Estimated relative probabilities of selection for different habitats by bighorn sheep with lower and upper simultaneous 90% confidence limits computed using the Bonferroni inequality with a confidence level of $(100-10/10)\% = 99\%$ for the ten individual intervals

Habitat type	u_{i+}	$\pi_i u_{++}$	$\hat{w}_i$	$se(\hat{w}_i)$	Bonferroni Confidence limits	
					Lower	Upper
Riparian	0	26.700	0.000	0.000	–	–
Conifer	6	57.850	0.104	0.037	0.009	0.198
Mt. shrub I	9	71.200	0.126	0.036	0.033	0.220
Aspen	18	66.750	0.270	0.081	0.061	0.479
Rock outcrop	14	26.700	0.524	0.213	0.000[*]	1.074
Sage/bitterbrush	63	75.650	0.833	0.211	0.289	1.376
Windblown ridges	46	53.400	0.861	0.105	0.590	1.133
Mt. shrub II	62	17.800	3.483	0.471	2.270	4.696
Prescribed Burns	178	40.050	4.444	0.407	3.397	5.492
Clearcut	49	8.900	5.506	1.717	1.082	9.929

[*]An impossible negative confidence limit for rock outcrop has been replaced by 0.000.

It is seen, for example, that the selection index for the prescribed burn is $\hat{w}_9 = 4.44$, with a confidence interval of 3.40 to 5.49 for the population value. Since the interval does not include the value 1, there is significant selection for this habitat above what would be expected by chance. Also, mountain shrub II and clearcut have relative probabilities of selection which are above that expected under the hypotheses of no selection. Similarly, there is significant selection against conifer, mountain shrub I, and aspen relative to the amount of habitat available. There is no apparent selection for or against rock outcrop, sage/bitterbrush, and windblown ridges. Since riparian was not used at all it seems clear that this was selected against.

Consider next the comparison of pairs of selection ratios using the confidence limits (4.34). There are 45 possible comparisons between pairs of these ratios, which suggests that a reasonable α value for the limits is $0.1/45 = 0.002$. In other words, the individual confidence limits should have 99.8% confidence in order that there is a probability of approximately 0.9 that all of the intervals will contain the population values. This is achieved by using $z_{\alpha/(2I)} = z_{0.001} = 3.09$ in the limits (4.34). For example, consider the comparison of the selection ratios for clearcut ($\hat{w}_{10} = 5.506$) and prescribed burns ($\hat{w}_9 = 4.444$). From equation (4.33), $se(\hat{w}_{10}-\hat{w}_9) = 2.01$, so that the confidence limits for $w_{10}-w_9$ are $1.062 - 3.09(2.01)$ to $1.062 + 3.09(2.01)$, which is -5.149 to 7.273. Since this includes zero, the difference between the use of clearcut and prescribed burns is not significant.

Table 4.9 shows which selection ratios are significantly different on this basis, when the habitats are listed in order of their estimated ratio. It can be seen that mountain shrub II and prescribed burns are selected with significantly higher probability than are lower ranking habitat types. The use of clearcut has a high variance because the fifth sheep never used this habitat but the sixth sheep used it extensively. Thus, even though clearcut has the largest estimated selection index of $\hat{w}_{10} = 5.506$, it is not significantly larger than the selection indices for some of the lower ranking habitats. If there had been no replication between sheep then the high variance between sheep would be hidden and incorrect statistical conclusions might be reached concerning the use of the clearcut habitat.

Table 4.9 Significant differences between estimated selection ratios calculated from the data in Table 4.7

Habitat	Rip.	Con.	Ms1.	Asp.	Roc.	S/b.	Wbr.	Ms2.	Prb.
Conifer	−								
Mt. shrub I	+	−							
Aspen	+	−	−						
Rock outcrop	−	−	−	−					
Sage/bitterbrush	+	−	+	−	−				
Windblown ridges	+	+	+	+	−	−			
Mt. shrub II	+	+	+	+	+	+	+		
Prescribed burns	+	+	+	+	+	+	+	−	
Clearcut	+	+	+	+	−	−	−	−	−

The entry '+' indicates a significant difference between the row habitat and the column habitat, with column habitats having obvious abbreviations for their names. The entry '−' indicates no significant difference.

Although the simulation study mentioned earlier did not address the question of the validity of confidence intervals for differences between selection ratios, it is possible to say something about this. Some limited simulations indicated that the statistics $z_{ik} = \{(\hat{w}_i-\hat{w}_k) - (w_i-w_k)\}/se(\hat{w}_i-\hat{w}_k)$ have distributions that are rather similar to the distributions of the statistics $z_i = (\hat{w}_i-w_i)/se(\hat{w}_i)$ as shown in Figure 4.2. Thus the tendency for confidence intervals for selection ratios to be conservative seems to be shared by confidence intervals for differences between selection ratios.

4.12 Sample of available resource units

If the proportion π_i of the ith resource category available has to be estimated then an additional source of variation is introduced into estimates of selection ratios. Thus, the selection ratio for the ith resource category by the jth animal is estimated by

$$\hat{w}_{ij} = u_{ij}/(\hat{\pi}_i \, u_{+j}), \tag{4.35}$$

where $\hat{\pi}_i$ is the estimated proportion of resource category i available. This is equation (4.28) but with π_i estimated.

In this situation we recommended that the estimator used for the selection ratio for resource units in category i is

$$\hat{w}_i = u_{i+}/(\hat{\pi}_i u_{++}), \tag{4.36}$$

which is equation (4.29) but with π_i estimated. The variance of this estimator can be determined in two stages. First, the variance of the ratio $V_i = u_{i+}/u_{++}$ can be estimated using equation (4.3), taking $y_j = u_{ij}$ and $x_j = u_{+j}$. Next, assuming that $\hat{\pi}_i$ is estimated from an independent source of data with standard error $se(\hat{\pi}_i)$, equation (4.4) can be used to estimate the variance of the final ratio $\hat{w}_i = V_i/\hat{\pi}_i$, setting $y = V_i$, $x = \hat{\pi}_i$, and $r_{xy} = 0$. Bonferroni simultaneous confidence intervals for a set of selection ratios can then be calculated as discussed in section 4.7.

To compare selection ratios the differences

$$\hat{w}_i - \hat{w}_k = u_{i+}/(\hat{\pi}_i \, u_{++}) - u_{k+}/(\hat{\pi}_k u_{++}) \tag{4.37}$$

can be calculated together with their standard errors, for all possible values of i and k. Bonferroni simultaneous confidence limits can then be determined as discussed in section 4.8.

Equation (4.37) has the same form as equation (4.5), with identification being established by setting $y_{1j} = u_{ij}$, $x_{1j} = \hat{\pi}_j u_{+j}$, $y_{2j} = u_{kj}$ and $x_{2j} = \hat{\pi}_k u_{+k}$. The variance of $\hat{w}_i - \hat{w}_k$ can therefore be estimated by equation (4.6), although a modified version of this equation taking into account how the available resource proportions are estimated may be more convenient to use.

Probably the most commonly used procedure involves choosing m_+ random resource units from the available population. For example, random points in a study area (or on a map of a study area) might be chosen and the habitat type encountered at each point recorded. The proportion of points encountering habitat type i, $\hat{\pi}_i$, is then taken as the estimate of π_i. With this method of estimation the numbers of resource units in different categories follow a multinomial distribution so that

$$var(\hat{\pi}_i) = \pi_i(1-\pi_i)/m_+,$$

and

$$cov(\hat{\pi}_i, \hat{\pi}_k) = -\pi_i \pi_k/m_+.$$

Using these variances and covariances, it can be shown that equation (4.6) yields the result

$$\hat{v}ar(\hat{w}_i-\hat{w}_k) = \sum_{j=1}^{n} \{u_{ij}/\hat{\pi}_i-u_{kj}/\hat{\pi}_k-(\hat{w}_i-\hat{w}_k)u_{+j}\}^2/(n-1)\{n/u_{++}^2\}$$

$$+ \{\hat{w}_i^2/\hat{\pi}_i +\hat{w}_k^2/\hat{\pi}_k - (w_i-\hat{w}_k)^2\}/m_+. \qquad (4.38)$$

Because of the nature of the equations that have been provided in this section for standard errors, it is difficult to be sure how accurate they will be in practice. This matter is considered in the following example, although at this point it can be said that it seems clear that it is particularly important to get good estimates of population proportions of available resource units in different categories since these play such a key role in the estimation procedures.

Chi-squared tests can be used to test for whether animals are using resources in a similar way, and whether they are being selective. Thus the statistic X_{L1}^2 of equation (4.26) provides a test for the consistent use of resources by different animals since it does not take availability into account. Also, if the availability of different resource categories is determined by a random sample of m_+ available resource units, then X_{L1}^2 can be calculated treating this sample as an (n+1)th 'animal'. This then gives a test for consistent proportions both for animals and the availability sample. A significant result indicates that selection occurs since this is the case either if the animals use resources categories with different probabilities or if their use is the same but differs from what is expected from the available sample.

4.12.1 Example 4.5 Habitat selection by bighorn sheep (partly artificial data)

For the sake of an example, suppose that Arnett et al. (1989) had not been able to determine the proportions of different habitat available for bighorn sheep exactly, but had instead estimated these proportions from a sample of 250 random points in the study region. They might then have obtained the data shown in Table 4.10.

Since the counts of used units for the six animals are the same here as in Table 4.7, the chi-squared statistic of equation (4.26) has the same value as was the case for example 4.4 where available proportions were assumed known. This is $X_{L1}^2 = 99.2$, with 45 degrees of freedom. It tests for a consistent choice of resources from animal to animal, irrespective of whether this is selective or not. As noted in example 4.4, this is highly significant and indicates a lack of consistency.

If the sample of available resources is included in the chi-squared calculation as a type of non-selective (n+1)th 'animal', then the chi-squared value increases to $X_{L2}^2 = 373.4$, with 54 degrees of freedom. The difference $X_{L2}^2-X_{L1}^2 = 274.2$ with nine degrees of freedom is then a measure of the amount of selectivity, irrespective of whether there are differences between animals or not. Again the result is highly significant, indicating very strong evidence of selection.

As mentioned in example 4.4, the small frequencies for some of the habitat categories, particularly the zero frequencies for 'riparian', mean that the accuracy of the chi-squared approximations is questionable for these data. However, the very large chi-squared values means that the evidence for selectivity and differences between animals is clearly established.

Estimates of selection ratios and their standard errors are shown in the final two columns of Table 4.10. These can be used to produce Bonferroni confidence intervals of the form

$$\hat{w}_i \pm z_{\alpha/(2 \times 10)} se(\hat{w}_i)$$

for the population selection ratios. The justification for these limits has been discussed before in this chapter. Here we simply note that all the ten possible intervals are expected to contain their respective population values with probability $1-\alpha$. Of course, the situation with the 'riparian' habitat which was not used at all by the sheep is unsatisfactory since the estimated selection ratio is zero, with estimated standard error of zero. This habitat could therefore reasonably be excluded from the confidence interval calculations.

Table 4.10   The bighorn sheep data with an artificial sample of 250 available resource units instead of known population proportions in different resource categories

Habitat	Available sample		Use by bighorn sheep						$\hat{w}$	$se(\hat{w})$
	Count	Proportion	1	2	3	4	5	6		
Riparian	21	0.084	0	0	0	0	0	0	0.00	0.00
Conifer	26	0.104	0	2	1	1	0	2	0.12	0.05
Mt. shrub I	41	0.164	0	1	2	3	2	1	0.12	0.03
Aspen	40	0.160	2	2	1	7	2	4	0.25	0.08
Rock outcrop	14	0.056	0	2	0	5	5	2	0.56	0.27
Sage/bitterbrush	46	0.184	16	5	14	3	18	7	0.76	0.21
Windblown ridges	34	0.136	5	10	9	6	10	6	0.76	0.15
Mt. shrub II	8	0.032	14	10	8	9	6	15	4.35	0.16
Prescribed burns	14	0.056	28	35	40	31	25	19	7.14	1.96
Clearcut	6	0.024	8	9	4	9	0	19	4.58	2.34
Total	250		73	76	79	74	68	75		

Bonferroni confidence intervals of the form

$$\hat{w}_i - \hat{w}_k \pm z_{\alpha/(2 \times 45)} se(w_i - \hat{w}_k)$$

can also be constructed for differences between population selection ratios using equation (4.38). Again, the probability that all of the 45 possible intervals contain their population values should be approximately $1-\alpha$.

The procedure for constructing and interpreting these limits is the same as was used for example 4.4, except for the calculation of standard errors. Details will therefore not be provided here. However, it is useful to summarize the results of a simulation study that was designed to assess the validity of the proposed confidence intervals for population selection ratios.

As for previous examples, it can be argued that the validity of confidence intervals of the form $\hat{w}_i \pm z_\alpha se(\hat{w}_i)$, where z_α is the value exceeded with probability α, depends on $z_i = (\hat{w}_i - w_i)/se(\hat{w}_i)$ having a standard normal distribution. A reasonable way to assess the limits therefore involves generating artificial data of a similar nature to real data, calculating z_i values, and comparing the distribution obtained for these with the standard normal distribution.

This exercise has been carried out using the data in Table 4.10 to provide expected frequencies for a sample of available resource units and samples from six 'sheep'. The actual data frequencies generated had Poisson distributions about the expected frequencies. A total of 225 sets of data were generated, each providing values for z_2 to z_{10}. Riparian was never used for the simulated sets of data since the Poisson expected frequency was set to zero for all six sheep for this habitat type. Hence z_1 was always undefined.

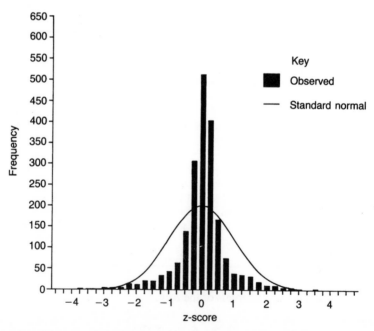

Figure 4.3 The distribution of values of $z_i = (\hat{w}_i - w_i)/se(\hat{w}_i)$ obtained by simulating 225 sets of data similar to the observed data that are shown in Table 4.10. It is required that z_i has a standard normal distribution for confidence limits for population selection ratios to be valid with data of the form shown in Table 4.10.

The distribution obtained for the $225 \times 9 = 2025$ generated z values is shown in Figure 4.3 together with the standard normal distribution. This figure shows a great similarity to Figure 4.2 since in both cases the distribution of z values is generally less variable than the standard normal. This then suggests that confidence intervals of the form $\hat{w}_i \pm z_\alpha se(\hat{w}_i)$ will generally be conservative, i.e., the limits will contain their population parameters with a higher probability than $1-\alpha$.

Although the validity of confidence limits for differences between selection ratios was not investigated in the simulations, it does seem likely that the conservative nature of the confidence limits for individual selection ratios will carry over since (as discussed in example 4.4) this is what seems to happen if the population proportions of available resource units of different types are known.

4.13 DESIGN III WITH SAMPLING PROTOCOL A

With design III studies the use and availability of resource units is measured separately for each animal. As for the other designs, there are then two cases to consider, depending on whether the proportions of units of different types that are available to each animal are known accurately, or are estimated from a random sample. We consider first the case where the available proportions are known accurately.

4.14 CENSUS OF AVAILABLE RESOURCE UNITS

We assume as before that there are I resource categories, the use of resources is measured for n animals, and that a random sample of u_{+j} resource units used by the jth animal is obtained and found to contain u_{ij} units in resource category i. We assume also that π_{ij} is the known proportion of resources available to animal j that are in category i.

An estimate of the selection ratio for the ith resource category by the jth animal is then

$$\hat{w}_{ij} = u_{ij}/\pi_{ij}u_{+j}. \tag{4.39}$$

Following our previous methods for combining results from different animals, this then suggests that the estimator to use for the selection ratio for the population of animals for the ith resource category is

$$\hat{w}_i = u_{i+}/\sum_{j=1}^{n} \pi_{ij}u_{+j}. \tag{4.40}$$

Equation (4.3) can be used to estimate the variance of $\hat{w}_i$ by setting $y_j = u_{ij}$ and $x_j = \pi_{ij}u_{+j}$, so that $\hat{R} = \hat{w}_i$. Bonferroni confidence intervals for population selection ratios can then proceed in the same way as for design I and design II.

To compare two estimated selection ratios the difference $\hat{w}_i - \hat{w}_k$ can be considered with an estimate of the standard error of this difference. To this end it can be noted that $\hat{w}_i-\hat{w}_k$ takes the form of equation (4.5) and that the use of equation (4.6) yields

$$\text{var}(\hat{w}_i-\hat{w}_k) = \{n/(n-1)\} \sum_{j=1}^{n} \{(u_{ij}-\hat{w}_i u_{+j})/(\pi_{i1}u_{+1} + ... + \pi_{in}u_{+n})$$
$$+ (u_{ik}-\hat{w}_k u_{+j})/(\pi_{k1}u_{+1} + ... + \pi_{kn}u_{+n})\}^2. \quad (4.41)$$

This variance is the same as what is given by equation (4.33) if $\pi_{ij} = \pi_i$ and $\pi_{kj} = \pi_k$ for all j, i.e., if all animals have the same available proportions for resource categories i and k.

Once the differences $\hat{w}_i-\hat{w}_k$ have been calculated together with their standard errors, a set of Bonferroni confidence limits can be calculated in such a way that there is a high probability that all of the population differences are within the limits. The procedure is exactly the same as for design I and design II.

A test for whether the jth animal sampled is selective is provided by calculating

$$X_{Lj}^2 = \sum_{i=1}^{I} u_{ij}\log_e\{u_{ij}/(u_{+j}\pi_{ij})\},$$

with I-1 degrees of freedom, where this compares the observed use of category i resource units (u_{ij}) with the expected use based on availability ($u_{+j}\pi_{ij}$). Adding up the n statistics obtained in this way then provides an overall test for selection with n(I-1) degrees of freedom. As usual, these chi-squared tests require the expected frequencies $u_{+j}\pi_{ij}$ to be 'large', which means that most if not all of them should be five or more.

4.15 SAMPLE OF AVAILABLE RESOURCE UNITS

The final type of study design that we will consider in detail is design III with the resources available to individual animals being estimated independently, for example by taking random samples of habitat points within home ranges.

Under these conditions equation (4.39) which applies when the resources available to individual animals are known, changes in the obvious way to

$$\hat{w}_{ij} = u_{ij}/\hat{\pi}_{ij}u_{+j}, \quad (4.42)$$

where $\hat{\pi}_{ij}$ is the estimated proportion of the resource units that are available to the jth animal that are in category i. Similarly in equation (4.40) the known available resource proportions are replaced by their estimates to give

$$\hat{w}_i = u_{i+}/\sum_{j=1}^{n} \hat{\pi}_{ij}u_{+j}. \quad (4.43)$$

To estimate the standard error of $\hat{w}_i$, equation (4.3) can be used yet again. It is simply necessary to set $y_j = u_{ij}$ and $x_j = (\hat{\pi}_{ij}u_{+j})$, so that $\hat{R} = \hat{w}_i$. Bonferroni confidence intervals can then be constructed in the usual way.

To compare selection ratios, all possible differences $\hat{w}_i - \hat{w}_k$ can be calculated together with estimated standard errors, and Bonferroni simultaneous confidence intervals constructed as for the other designs that have been considered. To this end, equation (4.7) can be used taking $y_{1j} = u_{ij}$, $x_{1j} = \hat{\pi}_{ij}u_{+j}$, $y_{2j} = u_{kj}$, $x_{2j} = \hat{\pi}_{kj}u_{+j}$.

4.16 DISCUSSION

In section 2.4 a number of general assumptions were mentioned as being required in order to estimate resource selection functions. In the context of the present chapter these assumptions are what is required for the valid estimation of selection ratios, and it is appropriate at this point to review the assumptions in this light.

Assumption (a) is that the proportions of different categories of resource units that are available do not change during the sampling period. For example, this assumption might be violated if animals eat most of the food in their 'preferred' habitat during the first two weeks of a four-week study. This requirement (a) is difficult to satisfy with many studies.

Assumption (b) is that the universe of available resource units is correctly identified. This may be difficult with design III studies because of the need to identify what is available to individual animals.

Assumption (c) is that the universe of used resource units is correctly identified and sampled. For example, this requirement may be violated if animals are eating food items which are not detectable in faecal samples. This is one of the most crucial and most difficult assumptions of the study design. Specific applications of the theory must be addressed separately and a general discussion would be unduly long. We note, however, that methods such as that developed by Nams (1989) for adjusting for radio telemetry errors may need to be used in some applications.

Assumption (d) is that the variables which actually influence the probability of selection are correctly identified. For example, this assumption is violated if percentage cover by vegetation is measured, but animals are actually selecting plots on the basis of the height of the plants. Hopefully, the variables under study will be highly correlated with the variables that actually influence the probability of selection.

Assumption (e) is that animals have unrestricted access to the entire distribution of available resource units. If animals are territorial then a few aggressive individuals may control all of the 'preferred' habitat, so that this assumption is violated. The assumption is most easily justified when the subpopulation of used units is small relative to the population of available units. Also, changes in the density of animals or in the availability of resource units may change the underlying selection strategies and the selection indices. Thus, statistical inferences are made with respect to the specific conditions present in the study area over the time period of interest.

Assumption (f) is that resource units are sampled randomly and independently. This requirement might be violated if sampled animals are in the same herd or if the visibility of animals varies with the habitat type. Estimates of selection indices may still be meaningful if this assumption is not satisfied, but standard errors may not reflect the true variation in the populations. For the sake of illustration, our example analyses were made on the assumption of random independent samples of resource units. It will be difficult to ensure this, especially in cases when animals occur in herds or when resource units are collected in batches. For example, consider the collection of stomach samples of animals in a design I food selection study. In this case the food items are obtained in batches and the selections of individual food items may not have been independent events because of different food preferences by different animals.

Another common but difficult situation is the analysis of relocations of radio-tagged animals. Relocations often come in a batch recorded at a series of points in time. Care must then be taken to ensure that the time interval between recordings is sufficient to assume that observations of used habitat points are independent events if the relocation points are to be considered the units of replication. See also Palomares and Delibes (1992).

In the presence of these problems, one approach is to estimate separate resource selection indices for several independent replications of batches of dependent units. Thus, one might estimate the selection indices for each of several randomly selected sites in a large study region. Inference toward mean values of the selection ratios over the entire study region can then proceed by standard statistical procedures, using replicates to determine standard errors. Alternatively, one might consider the selection of individual animals as independent events, and estimate separate selection indices for each animal by randomly sampling the units available and the units used by each animal to give what we have called a design II or a design III study. This may be the only reasonable approach for the study of food and habitat use by highly territorial animals or for the study of selection by radio-tagged animals.

In addition to the assumptions (a) to (f), we note that with design III studies estimates of the proportions of different types of resource units available (π_{ij}) may not be truly independent among animals. For example, a sample survey of habitat available in the overall study area may be conducted. Then the observations falling into an individual animal's home range might be used to estimate the habitat available to that specific animal. In that case, if there is considerable overlap of home ranges then some data points will influence the estimate of habitat availabilities for several different animals. At this time, the procedure described in section 4.15 is recommended for the estimation of variances of selection ratios. We note, however, that the true sampling variance of $\hat{w}_i$ may be underestimated. Of course, independent estimates of π_{ij} should be obtained for each animal if possible, so that estimates of sampling variances are approximately unbiased.

One possibility that we have not discussed involves using bootstrapping in an attempt to improve on the confidence intervals provided by normal distribution approximations. Also, it may be that the biases present in estimated selection ratios derived from small samples could be removed by modifying the estimators in minor ways. Unfortunately, the basic research needed to investigate these matters does not seem to have been carried out.

Exercise

This exercise concerns the use of five habitat types by grey partridges (*Perdix perdix*), as recorded by Smith *et al.* (1982). Ten grey partridges were radio-tagged and radio locations were classified in one of five habitats: small grain fields, row crop fields, hay fields, pasture or idle. Availability was censused by partitioning a map of the study area into five habitat types. A subset of the location data for one of six time periods is shown in Table 4.11, together with the percentages of different habitats in the study area.

Table 4.11 Numbers of radio locations in different habitat types for ten grey partridges, with the percentage of the available area in each habitat for the study region as a whole

	Habitat types					
Bird	Small grain fields	Row Crop	Hay	Pasture	Idle	Total
1	0	8	0	20	2	30
2	25	21	0	0	1	47
3	17	11	0	0	2	30
4	4	0	0	0	2	6
5	20	0	0	9	0	29
6	22	0	0	2	0	24
7	0	7	6	0	1	14
8	10	26	2	8	0	46
9	21	0	4	0	3	28
10	44	1	0	0	5	50
Total	163	74	12	39	16	304
Available (%)	28.2	41.7	10.2	13.5	6.3	100

Noting that this is a design II study with known proportions of different habitats available,

(a) test for evidence of selection using chi-squared tests and by constructing Bonferroni simultaneous confidence limits for population selection ratios; and

(b) construct Bonferroni simultaneous confidence limits for differences between the population selection ratios for different habitats.

5 Estimating a resource selection probability function from a census of resource units using logistic regression

One of the simplest ways of estimating a resource selection probability function involves taking a census of the used and unused units in a population of resource units, and estimating a logistic function for the probability of use as a function of variables that are measured on the units. This approach is discussed in this chapter, and is illustrated using data on the selection of winter habitat by antelope.

5.1 INTRODUCTION

Suppose that there are N available resource units that can be censused so that it is known which of these have been used and which have not been used after a single period of selection. The approach to data analysis that is proposed in this chapter involves the use of logistic regression to relate the probability of use to p X variables that are measured on the resource units, so that the resource selection probability function is assumed to take the form

$$w^*(x) = \frac{\exp(\beta_0 + \beta_1 x_1 + \ldots + \beta_p x_p)}{1 + \exp(\beta_0 + \beta_1 x_1 + \ldots + \beta_p x_p)}. \tag{5.1}$$

This function has the desirable property of restricting values of $w^*(x)$ to the range 0 to 1, but is otherwise arbitrary. Other functions that could be used include the probit

$$w^*(x) = \Phi(\beta_0 + \beta_1 x_1 + \ldots + \beta_p x_p), \tag{5.2}$$

where $\Phi(z)$ is the integral from $-\infty$ to z for the standard normal distribution, and the double exponential

$$w^*(x) = 1 - \exp\{-\exp(\beta_0 + \beta_1 x_1 + \ldots + \beta_p x_p)\}. \tag{5.3}$$

The main justification for using the logistic function rather than any other to approximate the resource selection probability function is the ready availability of computer programs for estimating this function. As will be discussed in Chapter 6, function (5.3) is more suitable for situations where the use of resource units is recorded for several selection periods involving different amounts of selection time.

5.2 ESTIMATING THE LOGISTIC FUNCTION

Suppose that the N available resource units can be divided into I groups so that within the ith group the units have the same values $x_i = (x_{i1}, x_{i2}, \ldots, x_{ip})$ for the X variables. The number of resource units used in group i, u_i, can then be assumed to be a random value from the binomial distribution with parameters A_i and $w^*(x_i)$, where A_i is the number of available resource units in the group, and maximum likelihood estimates of the β parameters in equation (5.1) can be calculated using any of the standard computer programs for logistic regression. The input that is required for estimation are the group sizes (A_1 to A_I), the vectors of X values (x_1 to x_I), and the numbers of used units for each group (u_1 to u_I).

Often all of the available resource units will have different values for the X variables, so that each of the K groups consists of just one resource unit. This causes no difficulties as far as estimation is concerned, and in fact some computer programs are specifically designed to handle this case only.

As discussed in section 3.3, minus twice the maximized log-likelihood for a fitted model can be used as a statistic indicating the goodness-of-fit of the model. In the present context this statistic is

$$X_L^2 = 2 \sum_{i=1}^{I} [u_i \log_e\{u_i/(A_i\hat{w}^*(x_i))\} + (A_i-u_i)\log_e\{(A_i-u_i)/(A_i-A_i\hat{w}^*(x_i))\}], \quad (5.4)$$

where the degrees of freedom are I–p–1. The condition for this to have an approximately chi-squared distribution is that most values of $A_i w^*(x)\{1-w^*(x)\}$ are 'large', which in practice means that they are five or more. However, differences between X_L^2 values for different models can reliably be tested against the chi-squared distribution even when the values of $A_i w^*(x)\{1-w^*(x)\}$ are small (McCullagh and Nelder, 1989, p. 119).

Some computer programs for logistic regression output the difference between the chi-squared statistics for the 'no selection' model with $w^*(x) = \exp(\beta_0)/\{1+\exp(\beta_0)\}$ and the particular model being fitted with one or more X variables included, but do not output the X_L^2 values themselves. It is therefore useful to note that the X_L^2 value for the 'no selection' model can be found by substituting $\hat{w}^*(x) = u_+/N$ in equation (5.4), where $u_+ = \Sigma u_i$ is the total number of used units out of the N available. The reason for this is that in the absence of selection the maximum likelihood estimate of the probability of use for all units is the observed proportion of units used. The 'no selection' goodness-of-fit value has I–1 degrees of freedom.

5.2.1 Example 5.1 Habitat selection by pronghorn

As an example of the use of logistic regression to assess resource selection, consider the study carried out by Ryder (1983) on winter habitat selection by pronghorn (*Antilocapra americana*) in the Red Rim area in south-central Wyoming that has already been described in example 2.2. Recall that Ryder set up 256 study plots and recorded the presence or absence of pronghorn in the winters of 1980–81 and 1981–82, together with a number of characteristics of each plot. The data are shown in Table 2.2 but with a number of vegetation height variables omitted since these are not defined on some study plots. Figure

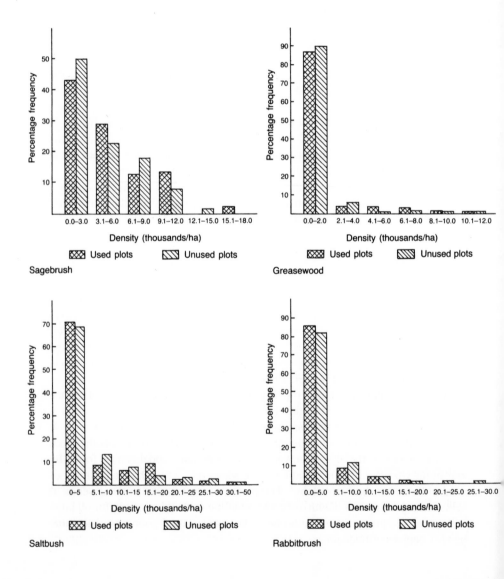

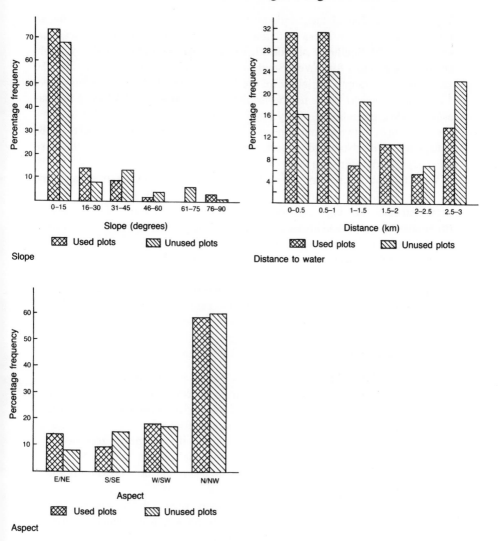

Figure 5.1 Distributions of the variables shown in Table 2.2 for 128 unused plots and 128 plots used at least once.

5.1 gives a comparison between the distributions of the variables for the unused plots and the plots used at least once. It can be seen from this figure that only the distribution of the distance to water is clearly different for these two groups.

To allow the estimation of a resource selection probability function where each of the four aspects (East/Northeast, South/Southeast, West/Southwest and North/Northwest) has a different probability of use, three 0–1 indicator variables can be used to replace the single aspect number shown in the last column of

Table 2.2. The first of these indicator variables can be set equal to 1 for an East/Northeast plot or otherwise 0, the second indicator variable can be set equal to 1 for a South/Southeast plot or otherwise 0, and the third indicator variable can be set equal to 1 for a West/Southwest plot or otherwise 0. For example, the values of these dummy variables for the first plot are 0 0 0 since this has aspect 4 (North/Northwest), while for the second plot the values are 0 0 1 since this has aspect 3 (West/Southwest). Only three indicator variables are needed to allow for differences between four aspects since the North/Northwest aspect can be considered as the 'standard' aspect, and it is only necessary to allow the three other aspects to differ from this.

With the introduction of the indicator variables for aspect there are nine variables available to characterize each of the 256 study plots: X_1 = density (thousands/ha) of big sagebrush (*Artemisia tridentata*); X_2 = density (thousands/ha) of black greasewood (*Sarcobatus vermiculatus*); X_3 = density (thousands/ha) of Nuttall's saltbush (*Atriplex nuttalli*); X_4 = density (thousands/ha) of Douglas rabbitbrush (*Chrysothamnus viscidiflorus*); X_5 = slope (degrees); X_6 = distance to water (m); X_7 = East/Northeast indicator variable; X_8 = South/Southeast indicator variable; and X_9 = West/Southwest indicator variable.

As noted in example 2.2, the fact that the study plots could be used once or twice during the study period means there are several approaches that can be used for analysing the data by logistic regression, depending on what definition of 'use' is applied. Here the obvious possibilities are:

(a) A study plot can be considered to be 'used' if antelope are recorded in either the first or the second winter (as in the comparisons made in Figure 5.1). On this basis the application of logistic regression to approximate the probability of a plot being used is straightforward.
(b) A study plot can be considered to be 'used' if antelope are recorded in both years. This leads to probabilities of 'use' that are smaller than for definition (a), but an analysis of the data using logistic regression is still straightforward.
(c) A study plot can be considered to be 'used' when antelope are recorded for the first time. This turns the situation into one where units are 'used up' since the pool of unused study plots decreases with time. This approach requires that the effect of time be modelled, which therefore means that logistic regression is not a convenient approach for data analysis. A method of analysis that allows for the effect of time is discussed in the next chapter.
(d) The two years can be considered to be replicates, in which case it is interesting to know whether the nature of the resource selection (if any) was different for the two years. In this case, logistic regression can be used separately in each year, or one equation can be fitted to both years of data. This leads to a more complicated analysis than is needed if one of the definitions (a) and (b) is used but better use is made of the data.

It is approach (d) that will be used for this example. Thus the observational unit will be taken to be a study plot in one year. There are 512 such units, each of which is recorded as either being used or not used. The question of whether it is reasonable to regard the two years as providing independent data is discussed further below. Here it will merely be noted that an examination of this assumption is required in order to establish the validity of the analysis being used.

Three logistic regression models have been fitted to the data. For model 1 it was assumed that the resource selection probability function was different for the two winters 1980–81 and 1981–82. Hence the logistic equation (5.1) was fitted separately to the data for each of the two years, with all the variables X_1 to X_9 included. The calculations were done using the logistic regression module from the SOLO package (BMDP, 1988) which produced the estimates with standard errors that are shown in Table 5.1.

Table 5.1   Results from fitting the logistic regression equation separately to the data on habitat selection by antelopes in 1980–81 and 1981–82

	1980–81			*1981–82*		
Variable	*Coefficient*	*Standard error*	*p value*	*Coefficient*	*Standard error*	*p value**
Constant	−0.896	0.410	0.029	−0.056	0.376	0.882
Sagebrush	0.015	0.044	0.727	0.045	0.041	0.267
Greasewood	0.057	0.073	0.433	−0.038	0.073	0.607
Saltbush	0.020	0.021	0.343	−0.001	0.020	0.967
Rabbitbrush	−0.021	0.046	0.642	−0.086	0.045	0.058
Slope	−0.0043	0.0082	0.603	−0.0043	0.0075	0.565
Distance to water	−0.00035	0.00018	0.051	−0.00031	0.00016	0.054
E/NE aspect	1.003	0.443	0.013	0.534	0.427	0.211
S/SE aspect	0.007	0.519	0.989	−0.068	0.443	0.878
W/SW aspect	0.714	0.393	0.069	−0.332	0.387	0.392

*The p values shown are obtained by calculating the ratios of estimates to their standard errors and finding the probability of a value that far from zero for a standard normal variable.

Chi-squared tests can be used to assess whether there is any evidence that the probability of use of a study plot was related to one or more of the variables being considered. This is done by seeing whether the log-likelihood chi-squared goodness-of-fit statistics obtained from fitting model 1 are significantly less than the log-likelihood chi-squared values for the 'no selection' model. In SOLO output the log-likelihood chi-squared value for the 'no selection' model minus the log-likelihood chi-squared value for a model with one or more X variables is called the 'model chi-squared'. However, other computer programs for logistic regression might label this statistic differently.

For 1980–81 the model chi-squared obtained from fitting the logistic regression including all X variables is 17.86 with nine degrees of freedom, which is significantly large at the 5% level. The equivalent statistic for 1981–82 is 12.98 with nine degrees of freedom, which is not significantly large at the 5% level. There is therefore some evidence of selection in 1980–81 but not in 1981–82. The sum of the two chi-squared values is 30.84 with 18 degrees of freedom. This is a measure of the evidence of selection for both years combined which is significantly large at the 5% level.

Inspection of the coefficients in Table 5.1 indicates that there is not much evidence that habitat selection was related to the vegetation variables or the slope in either 1980–81 or 1981–82. However, the coefficient for the distance to water is nearly significant at the 5% level in both years, and the East/Northeast dummy variable for aspect is significantly large at about the 1% level for the first year. It was therefore decided to refit the logistic regression equations, again separately for each year, with the vegetation variables and the slope omitted. This resulted in the estimates shown in Table 5.2 for what will be called model 2. Again, SOLO was used for the calculations.

Table 5.2 Results from fitting the logistic regression equation separately to the data on habitat selection by antelope in 1980–81 and 1981–82, with some variables omitted

Variable	1980–81			1981–82		
	Coefficient	Standard error	p value	Coefficient	Standard error	p value[*]
Constant	–0.655	0.256	0.011	–0.164	0.238	0.490
Distance to water	–0.00044	0.00017	0.010	–0.00033	0.00015	0.030
E/NE aspect	1.036	0.432	0.017	0.561	0.418	0.180
S/SE aspect	–0.086	0.504	0.865	–0.043	0.430	0.921
W/SW aspect	0.613	0.371	0.099	–0.216	0.367	0.555

[*]The p values are explained in Table 5.1.

The model chi-squared statistic for model 2 is 14.89 with four degrees of freedom for 1980–81, and 7.81 with four degrees of freedom for 1981–82. The first of these two values is significantly large at the 1% level and the second is significantly large at the 10% level. The total of 22.70 with eight degrees of freedom can be compared with the total of the model chi-squared values for model 1 of 30.75 with 18 degrees of freedom. The difference is 10.05 with ten degrees of freedom. Since this is not at all significantly large the simpler model 2 is preferable to model 1.

The somewhat similar coefficients for the two years that are shown in Table 5.2 suggest that it may be possible to get about as good a result by fitting all the data together with a dummy variable introduced to allow for a difference between the years. This produces what will be called model 3. The results of fitting this model using SOLO are shown in Table 5.3. The dummy variable 'Year' was set equal to 0 for all the 1980–81 results and 1 for all the 1981–82 results.

Table 5.3 Coefficients for a model fitted to the combined data for 1980–81 and 1981–82

Variable	Coefficient	Standard error	p value*
Constant	−0.613	0.199	0.002
Year	0.410	0.194	0.035
Distance to water	−0.00037	0.00010	0.001
E/NE aspect	0.786	0.301	0.009
S/SE aspect	−0.059	0.325	0.860
W/SW aspect	0.180	0.260	0.489

*The p values are explained in Table 5.1.

The model chi-squared value is 23.87, with five degrees of freedom, which is significantly large at the 0.1% level. The difference between this value and the model chi-squared for model 2 is $23.87 - 22.70 = 1.17$, with three degrees of freedom. This is not at all significantly large, indicating that model 3 is about as good as the more complicated model 2.

Looking at the results in Table 5.3, it can be seen that the coefficients of year, distance to water, and the dummy variable for the East/Northeast aspect are all significantly different from zero at the 5% level. The non-significant coefficients for the other two dummy variables for aspect merely indicate that the probabilities of use for the South/Southeast and West/Southwest aspects are about the same as the probabilities for the standard North/Northwest aspect.

The chi-squared tests that have been described for comparing model 1, model 2 and model 3 are summarised in Table 5.4 in an analysis of deviance table (since the chi-squared statistics are sometimes called 'deviances'). Models are listed from the simplest ('no selection') to the most complicated (model 1, with all nine variables used and different coefficients estimated for each year).

The goodness-of-fit statistic X_L^2 for the 'no selection' model is not output by SOLO. However, it can be found from equation (5.4) by substituting in $\hat{w}^*(x) = 166/512$, on the grounds that 166 of the 512 study units were used, so that this ratio is the maximum likelihood estimate of the probability of use of a unit when this probability is the same for all units. On this basis, X_L^2 for the 'no selection' model is found to be 645.12, with 511 degrees of freedom. Once this is known, the goodness-of-fit statistics for the other models that are shown in Table 5.4 can be found by adding on the appropriate model chi-squared values.

The fact that all the goodness-of-fit statistics shown in Table 5.4 are significantly large might be thought to show that none of the models is a satisfactory fit to the data. However, this is not the case since the condition for these statistics to have approximately chi-squared distributions (most values of $A_i w^*(x)\{1-w^*(x_i)\}$ being five or more) is certainly not met. Hence the chi-square approximation is not reliable for testing the goodness-of-fit statistics, although it can be used for testing differences between these statistics for different models.

Table 5.4 Analysis of deviance table for assessing the goodness-of-fit of different models for the antelope data

Model	Chi-squared goodness-of-fit statistic	Degrees of freedom	Model chi-squared (difference)	Degrees of freedom
No selection	645.12*	511		
			23.87[†]	5
Model 3: selection on distance to water and aspect, with dummy variable for year	621.26*	506		
			1.17	3
Model 2: selection on distance to water and aspect, with coefficients different in each year	620.09*	503		
			10.05	10
Model 1: selection on all variables in Table 5.1, with coefficients different in each year	610.04*	493		

*Significantly large at the 0.1% level when compared with critical values of the chi-squared distribution (chi-squared approximation is not reliable).
[†]Significantly large at the 0.1% level when compared with critical values of the chi-squared distribution (chi-squared approximation is reliable).

The amount of selection is indicated by Figure 5.2, which shows values of the estimated resource selection probability function

$$\hat{w}^*(x) = \frac{\exp\{-0.613+0.410(\text{YEAR})-0.00037(\text{DW})+0.786(\text{E/NE})-0.059(\text{S/SE})+0.180(\text{W/SW})\}}{1+\exp\{-0.613+0.410(\text{YEAR})-0.00037(\text{DW})+0.786(\text{E/NE})-0.059(\text{S/SE})+0.180(\text{W/SW})\}},$$

where YEAR indicates the 0–1 variable for the year, DW indicates the distance to water, and E/NE, S/SE and W/SW are the dummy variables for aspect. The probabilities of use calculated from this function are plotted against the distance from water, separately for the 1980–81 and 1981–82 winters, and the four aspects. There was apparently a maximum probability of use of about 0.65 for East/Northeast study plots close to water in 1980–81, and a minimum probability of use of about 0.20 for South/Southeast plots far from water in 1981–82.

Residual plots can be examined to see whether there are any systematic deviations between the data and model 3. However, the standardized residuals

$$R_i = \{u_i - A_i w^*(x_i)\}/\sqrt{[A_i w^*(x_i)\{1 - w^*(x_i)\}]} \tag{5.5}$$

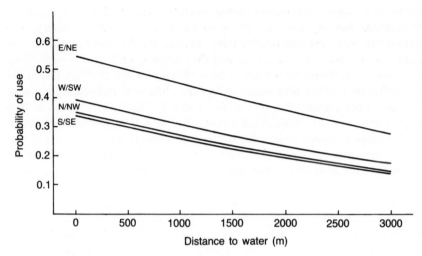

1980–81 winter

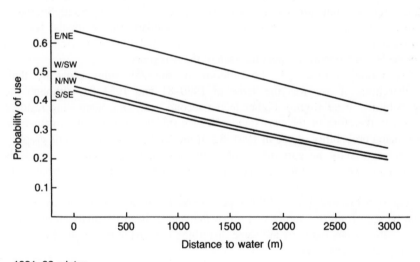

1981–82 winter

Figure 5.2 Probabilities of study plots being used according to the fitted model 3 logistic regression equation.

are not very informative since all the group sizes A_i are one. It is therefore unreasonable to expect these residuals to be approximately standard normally distributed. This problem can be overcome by grouping observations, and plotting standardized residuals for groups.

According to model 3, the probability of a study plot being used depended on the aspect, the distance to water, and the year. These are therefore the factors that the grouping of observations should depend on. Having decided this, it must be admitted that any basis for grouping has to be somewhat arbitrary. The method that was used here involved first dividing the 256 study plots into four groups on the basis of their aspect, and then ordering them within each group from those most distant from water to those closest to water. In effect, this meant that within each of the four aspect groups the plots were ordered according to their estimated probabilities of use for model 3. The study plots were then divided into sets of five within each aspect group, so that the first set consisted of the five plots estimated to be least likely to be used, followed by a set of five plots with higher estimated probabilities of use, and so on, with the last set allowed to have more or less than five plots in order to make use of all the plots available.

At this point, equation (5.5) was used to calculate two standardized residuals for each set of study plots within each aspect group, using average values for the probabilities of use $w^*(x_i)$. The first of the standardized residuals was based on the plots used in 1980–81, and the second one was based on the plots used in 1981–82. In this way, six standardized residuals were obtained for the East/Northeast aspect in 1980–81 and another six for this aspect in 1981–82. Similarly, six standardized residuals were calculated for the South/Southeast aspect for each of the two years, nine standardized residuals for the West/Southwest aspect for each of the two years, and 30 standardized residuals for the North/Northwest aspect for each of the two years.

According to model 3, the ordering of the 256 study plots by their probabilities of use was the same in 1980–81 and 1981–82, although the probabilities were slightly higher in the second year. One interesting residual graph is therefore of the two standardized residuals for each plot against the estimated probability of use in 1981–82. If model 3 is correct then this graph is expected to show no patterns at all, with most of the standardized residuals within a range from –2 to +2, and almost all of them within a range from –3 to +3.

Figure 5.3 shows graphs of this type, separately for each of the four aspects. It can be seen from this figure that all of the standardized residuals are within a reasonable range, but there are some disturbing patterns in the graphs for two of the aspects. In particular:

(a) For the East/Northeast aspect there are only six sets of study plots but the residuals are so similar for 1980–81 and 1981–82 that the assumption of independent data in the two years looks suspect. The Pearson correlation between the residuals for the two years is quite high at 0.48, although this is not significantly different from zero at the 5% level.

(b) The graph for the South/Southeast plots also indicates that the data are not independent for the two years. In this case the Pearson correlation is 0.85, which is significantly different from zero even with only six pairs of standardized residuals.

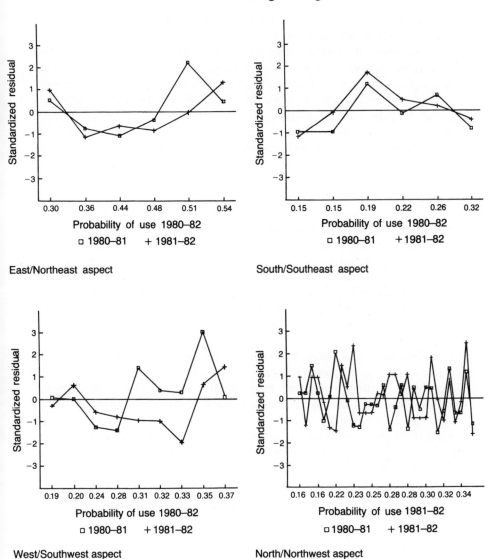

Figure 5.3 Standardized residuals plotted against the estimated probability of use in 1981–82, separately for each of the four aspects.

If the patterns for the first two aspects were repeated for the other two aspects as well then there would be little doubt that the assumption of independent data for the two years is untenable. However, the graphs for West/Southwest and North/Northwest study plots show little indication of dependence between the data for the two years, with the Pearson correlation coefficients being 0.22 based on nine pairs of standardized residuals, and 0.09 based on 30 pairs of standardized residuals, respectively.

Taken overall the residual plots do not show clear evidence of dependence between the data for 1980–81 and 1981–82 because the correlation between the standardized residuals for East/Northeast and South/Southeast plots is obscured by the low correlations for the other two aspects. In fact the Pearson correlation for all 51 pairs of standardized residuals is 0.20, which is not significantly different from zero at the 5% level. Still, there is cause for some concern and it is appropriate to conclude this example with a brief discussion of alternative explanations for the correlations indicated for the East/Northeast and South/Southeast residuals.

One explanation is, of course, that the behaviour of pronghorn is consistent from year to year so that an individual animal tends to be seen in the same study plots every year. If this is true then the data obtained for different years will be correlated, with the result that residuals will also be correlated. If this is the situation then the estimated resource selection probability function may still provide good estimates of probabilities of use for different plots. However, the calculated standard errors of β estimates will be too small since the effective amount of data available is less than the apparent amount. Also, the significance of differences in chi-squared values for the fits of different models will be exaggerated.

An alternative explanation for residuals from different years being correlated is that one or more important variables is missing from the resource selection probability function. In this case, the probability of use will be underestimated for some study plots and overestimated in others, and this bias will be present in both years. Consequently, there may be some plots with a high probability of use that are estimated to have a low probability of use. These plots will tend to be used in both years and hence provide positive residuals in both years. On the other hand, study plots with a low probability of use that are estimated to have a high probability of use will tend to give negative residuals in both years.

If this second explanation for correlated residuals is correct then there is a problem because estimated probabilities of use may be seriously biased. In the case of the present example there is nothing that can be done about this without taking further measurements on the plots of land. However, the graphs in figure 5.3 suggest that if there is a missing variable then the values of this variable are related to the distance to water in the East/Northeast study plots and in the South/Southeast study plots in much the same way as the standardized residuals, but show little relationship to the distance to water for study plots with the other two aspects.

5.3 DISCUSSION

The logistic regression model described in this chapter has much to recommend it when census data are available and there is no need to take into account varying amounts of selection time since this type of model is used widely in other applications and many computer programs for estimation exist. It can be used in particular for analysing data from:

(a) one universe of resource units censused before and after a single period of selection;
(b) two or more replicate universes of resource units censused before and after the same amount of selection time; or
(c) (as in the case of the example on habitat selection by antelope) replications of the selection process on a single universe of resource units, providing that the selection time is the same for each replication.

Although the only example that has been considered is a design I study in the terminology of Chapter 1, with resource availability and use being measured at the population level, this does not mean that logistic regression cannot be used with the other two designs. The antelope study would have had design II if the use of study plots by individual animals had been recorded. In principle, it would then have been possible to estimate a resource selection probability function for each animal using logistic regression. An interesting question would then be whether a model that allows each animal to have a different resource selection probability function gives a significantly better fit to the data than a model that assumes all animals have the same function. In a similar way, differences between sexes, age groups, etc. could be studied.

With a design III study, availability is measured for each animal as well as use. This again would permit a separate resource selection probability function to be estimated for each animal, and tests for differences between animals or groups of animals would be possible.

With either a design II or design III study it would be desirable to have enough animals to use differences between them to assess the accuracy of estimated resource selection probability functions, rather than relying on the standard errors produced by computer programs for logistic regression. Thus, if equation (5.1) is estimated using separate data for n animals, then the standard error of $\hat{\beta}_i$ can be estimated with n−1 degrees of freedom using the observed standard deviation of the n individual estimates. In this way, the estimation of a resource selection probability function for each animal gives a first stage analysis, and inferences concerning the population of animals can be carried out using second-stage analyses.

As noted in Chapter 1, the advantage of this approach is that it is still valid even if the observations on each animal are not independent, providing that different animals do give independent observations. If different animals do not give independent observations then it may still be possible to isolate independent groups of animals and conduct a first stage analysis on each of these groups. In that case, second stage analyses can be based on regarding the groups as providing replicates.

Finally, it can be mentioned that although the calculations for the examples in this chapter were carried out using SOLO, they can be done equally well using the computer program RSF that is described in the Preface.

Exercise

Many of the problems facing the biologist in studying resource selection by animals are also found by the archaeologist studying the use of resources by human societies. One such study concerns the location of prehistoric Maya sites within the Corozal District of Belize in Central America. The investigator was Green (1973) who discusses the proposition that 'sites were located so as to minimize the effort expended in acquiring scarce resources'.

The resource units being considered are plots of land. The whole study area was divided into 151 of these, each being a square with 2.5 km sides. Thirteen variables were measured on each square, related to soil types, vegetation types, distance to navigable water, the distance to Santa Rita (a possible prehistoric commercial and political centre), and the number of sites in neighbouring squares. One or two sites were known to exist on 29 of the squares, giving 34 sites in total.

The data for a selection of the variables measured by Green are shown in Table 5.5. Use logistic regression to see whether the presence of one or more sites on a square can be related to the measured characteristics. A point to note with this example is that the existence of some misclassification has to be accepted since Maya sites that have not been found may well exist on some of the squares that are recorded as being unused. Thus the estimated probability of a square being used will in fact be an estimate of the probability of use multiplied by the probability of a site being discovered. This need be of no concern providing that the probability of a site being discovered is approximately constant for all the existing sites. Exactly the same problem occurs with Ryder's study of habitat selection by antelope where the classification of a plot of land as used depends on an antelope being sighted at least once on that land.

Table 5.5 Data on the presence of prehistoric Maya sites in the Corozal District of Belize in Central America[*]

	Number of sites	Soil percentages				Vegetation percentages				Other variables			
		X_1	X_2	X_3	X_4	X_5	X_6	X_7	X_8	X_9	X_{10}	X_{11}	X_{12}
1	0	40	30	0	30	0	25	0	0	1	0.5	30	15.0
2	0	20	0	0	10	10	90	0	0	2	0.5	50	13.0
3	0	5	0	0	50	20	50	0	0	2	0.5	40	12.5
4	0	30	0	0	30	0	60	0	0	1	0.0	40	10.0

Table 5.5   (Cont.)

Number of sites		Soil percentages				Vegetation percentages				Other variables			
		X_1	X_2	X_3	X_4	X_5	X_6	X_7	X_8	X_9	X_{10}	X_{11}	X_{12}
5	0	40	20	0	20	0	95	0	0	3	1.3	30	13.8
6	0	60	20	0	5	0	100	0	0	4	2.8	0	11.5
7	0	90	0	0	10	0	100	0	0	3	2.5	0	9.0
8	0	100	0	0	0	20	80	0	0	3	2.5	0	7.5
9	0	0	0	0	10	40	60	0	0	2	1.3	50	8.8
10	2	15	0	0	20	25	10	0	0	0	0.0	50	9.0
11	0	20	0	0	10	5	50	0	0	1	0.5	40	10.0
12	0	0	0	0	50	5	60	0	0	1	0.5	50	11.0
13	0	10	0	0	30	30	60	0	0	2	3.8	20	7.0
14	0	40	0	0	20	50	10	0	0	1	2.3	50	7.0
15	0	10	0	0	40	80	20	0	0	1	3.0	0	7.5
16	0	60	0	0	0	100	0	0	0	0	3.0	0	8.8
17	0	45	0	0	0	5	60	0	0	0	0.3	45	12.5
18	0	100	0	0	0	100	0	0	0	0	2.0	45	10.3
19	1	20	0	0	0	20	0	0	0	0	0.0	100	12.5
20	0	0	0	0	60	0	50	0	0	0	0.3	50	15.0
21	0	0	0	0	80	0	75	0	0	0	0.5	50	14.8
22	0	0	0	0	50	0	50	0	0	0	0.0	50	16.3
23	0	30	10	0	60	0	100	0	0	2	2.5	20	14.8
24	0	0	0	0	50	0	50	0	0	0	0.0	50	16.5
25	0	50	20	0	30	0	100	0	0	3	2.5	0	15.0
26	0	5	15	0	80	0	100	0	0	1	2.5	0	12.5
27	0	60	40	0	0	10	90	0	0	2	4.0	0	10.0
28	0	60	40	0	0	50	50	0	0	2	7.8	0	7.5
29	0	94	5	0	0	90	10	0	0	2	10.0	0	6.3
30	0	80	0	0	20	0	100	0	0	1	3.0	0	11.0
31	0	50	50	0	0	25	75	0	0	3	5.2	0	9.8
32	0	10	40	50	0	75	25	0	0	3	7.5	0	6.5
33	0	12	12	75	0	10	90	0	0	2	5.3	0	4.0
34	0	50	50	0	0	15	85	0	0	2	5.0	0	11.3
35	1	50	40	10	0	80	20	0	0	3	7.3	0	9.8
36	0	0	0	100	0	100	0	0	0	0	7.0	0	6.3
37	0	0	0	100	0	100	0	0	0	0	3.8	0	4.8
38	0	70	30	0	0	50	50	0	0	2	4.5	0	11.5
39	0	40	40	20	0	50	50	0	0	2	8.8	0	10.0
40	0	0	0	100	0	100	0	0	0	0	6.3	0	7.5
41	1	25	25	50	0	100	0	0	0	1	3.8	0	5.2
42	0	40	40	0	20	80	20	0	0	3	2.0	0	4.0
43	0	90	0	0	10	100	0	0	0	1	5.0	0	3.8
44	0	100	0	0	0	100	0	0	0	0	3.8	0	5.0
45	0	100	0	0	0	90	10	0	0	0	2.5	25	7.6

Table 5.5   (Cont.)

Number of sites		Soil percentages				Vegetation percentages				Other variables			
		X_1	X_2	X_3	X_4	X_5	X_6	X_7	X_8	X_9	X_{10}	X_{11}	X_{12}
46	1	10	0	0	90	100	0	0	0	2	3.5	0	2.5
47	1	80	0	0	20	100	0	0	0	1	2.8	5	0.0
48	0	60	0	0	30	80	0	0	0	1	1.3	50	3.0
49	0	40	0	0	0	0	30	0	0	0	0.0	100	5.3
50	2	50	0	0	50	100	0	0	0	1	2.0	50	2.0
51	2	50	0	0	0	40	0	0	0	0	0.0	100	1.3
52	1	30	30	0	20	30	60	0	0	2	1.3	50	4.0
53	0	20	20	0	40	0	100	0	0	2	1.0	50	17.6
54	0	20	80	0	0	0	100	0	0	1	3.0	0	15.2
55	0	0	10	0	60	0	75	0	0	1	0.3	50	21.3
56	0	0	50	0	30	0	75	0	0	2	2.8	20	18.8
57	0	50	50	0	0	30	70	0	0	2	5.5	80	16.3
58	0	0	0	0	60	0	60	0	0	0	0.0	50	24.0
59	0	20	20	0	60	0	100	0	0	2	2.5	20	21.5
60	1	90	10	0	0	70	30	0	0	1	5.0	0	20.0
61	0	100	0	0	0	100	0	0	0	0	6.3	0	17.6
62	0	15	15	0	30	0	40	0	0	2	1.0	50	25.2
63	1	100	0	0	0	25	75	0	0	0	3.0	0	23.8
64	1	95	0	0	5	90	10	0	0	0	5.5	0	21.4
65	0	95	0	0	5	90	10	0	0	0	8.0	0	20.0
66	1	60	40	0	0	50	50	0	0	1	6.0	0	12.6
67	0	30	60	10	10	50	40	0	0	3	8.5	0	11.0
68	1	50	0	50	50	100	0	0	0	1	3.0	0	9.0
69	1	60	30	0	10	60	40	0	0	1	1.3	25	7.5
70	1	90	8	0	2	80	20	0	0	1	7.5	0	14.8
71	1	30	30	30	40	60	40	0	0	4	4.8	0	11.5
72	1	33	33	33	33	75	25	0	0	3	1.8	40	11.0
73	0	20	10	0	40	0	100	0	0	2	0.0	100	9.8
74	0	50	0	0	50	40	60	0	0	1	5.3	0	16.0
75	0	75	12	0	12	50	50	0	0	2	2.5	0	14.8
76	0	75	0	0	25	40	60	0	0	1	0.5	100	13.0
77	0	30	0	0	50	0	100	0	0	2	0.0	100	11.5
78	0	50	10	0	30	5	95	0	0	3	5.0	0	17.5
79	0	100	0	0	0	60	40	0	0	1	2.5	0	17.3
80	0	50	0	0	50	20	80	0	0	2	0.0	100	15.0
81	0	10	0	0	90	0	100	0	0	1	0.3	100	14.9
82	0	30	30	0	20	0	85	0	0	3	0.8	80	6.3
83	0	20	20	0	20	0	75	0	0	3	0.0	100	6.3
84	1	90	0	0	0	50	25	0	0	0	0.5	100	7.5
85	0	30	0	0	0	30	5	0	0	0	0.0	100	8.7
86	2	20	30	0	50	20	80	0	0	4	1.0	100	8.8

Table 5.5   (cont.)

Number of sites		Soil percentages				Vegetation percentages				Other variables			
		X_1	X_2	X_3	X_4	X_5	X_6	X_7	X_8	X_9	X_{10}	X_{11}	X_{12}
87	0	50	30	0	10	50	50	0	0	1	0.0	100	8.8
88	0	80	0	0	0	70	10	0	0	0	1.8	100	8.9
89	1	80	0	0	0	50	0	0	0	0	0.8	100	10.0
90	0	60	10	0	25	80	15	0	0	3	1.3	50	11.3
91	0	50	0	0	0	75	0	0	0	0	0.0	100	11.3
92	0	70	0	0	0	75	0	0	0	0	0.0	100	11.5
93	0	100	0	0	0	85	15	0	0	0	2.5	0	13.3
94	0	60	30	0	0	40	60	0	0	3	2.5	25	13.3
95	0	80	20	0	0	50	50	0	0	1	0.0	100	13.8
96	0	100	0	0	0	100	0	0	0	0	2.5	40	14.5
97	0	100	0	0	0	95	5	0	0	0	5.0	0	15.0
98	0	0	0	0	60	0	50	0	0	2	0.3	45	34.0
99	0	30	20	0	30	0	60	0	40	3	1.3	45	32.5
100	0	15	0	0	35	20	30	0	0	0	0.0	50	40.0
101	1	40	0	0	45	70	20	0	0	2	1.3	50	37.8
102	0	30	0	0	45	20	40	0	20	3	0.0	100	35.2
103	0	60	10	0	30	10	65	5	20	3	1.3	20	33.8
104	0	40	20	0	40	0	25	0	75	3	1.0	60	27.0
105	1	100	0	0	0	70	0	0	30	0	3.0	0	25.0
106	1	100	0	0	0	40	60	0	0	2	6.0	0	23.5
107	0	80	10	0	10	40	60	0	0	2	8.0	0	21.4
108	1	90	0	0	10	10	0	0	90	0	1.3	75	28.8
109	1	100	0	0	0	20	10	0	70	0	3.0	0	26.5
110	0	30	50	0	20	10	90	0	0	2	6.0	0	25.0
111	0	60	40	0	0	50	50	0	0	1	5.3	0	23.3
112	0	100	0	0	0	80	10	0	10	0	2.5	0	33.0
113	1	60	0	0	40	60	10	30	0	1	4.8	0	28.4
114	0	50	50	0	0	0	100	0	0	2	7.0	0	27.0
115	0	60	30	0	10	25	75	0	0	3	4.5	0	25.5
116	0	40	0	0	60	30	20	50	0	1	5.0	0	31.5
117	0	30	0	0	70	0	50	50	0	2	7.5	0	30.3
118	0	50	20	0	30	0	100	0	0	3	6.0	0	29.0
119	0	50	50	0	0	25	75	0	0	1	6.5	0	27.5
120	0	90	10	0	0	50	50	0	0	1	5.5	0	20.2
121	0	100	0	0	0	60	40	0	0	0	3.0	0	18.5
122	0	50	0	0	50	70	30	0	0	1	0.0	100	17.5
123	0	10	10	0	80	0	100	0	0	2	0.3	100	17.4
124	0	50	50	0	0	30	70	0	0	2	3.8	0	22.0
125	1	75	0	0	25	80	20	0	0	1	1.3	90	20.5
126	0	40	0	0	60	0	100	0	0	2	0.3	90	20.0
127	0	90	10	0	10	75	25	0	0	2	3.5	20	19.0

Table 5.5   (Cont.)

Number of sites		X_1	X_2	X_3	X_4	X_5	X_6	X_7	X_8	X_9	X_{10}	X_{11}	X_{12}
		Soil percentages				Vegetation percentages				Other variables			
128	0	45	45	0	55	30	70	0	0	2	2.3	30	23.8
129	0	20	35	0	80	10	90	0	0	2	0.3	100	22.8
130	0	80	0	0	20	70	30	0	0	2	2.8	10	22.3
131	0	100	0	0	0	90	0	0	0	0	5.0	0	21.3
132	0	75	0	0	25	50	50	0	0	2	1.0	60	26.3
133	0	60	5	0	40	50	50	0	0	2	0.3	100	25.0
134	0	40	0	0	60	60	40	0	0	1	2.8	0	24.0
135	0	60	0	0	40	70	15	0	0	1	5.0	0	23.8
136	0	90	10	0	10	75	25	0	0	1	2.0	30	16.3
137	0	50	0	5	0	30	20	0	0	0	0.0	100	16.3
138	0	70	0	30	0	70	30	0	0	1	2.0	20	17.0
139	0	60	0	40	0	100	0	0	0	1	4.8	0	17.5
140	2	50	0	0	0	50	0	0	0	0	0.0	100	19.0
141	0	30	0	50	0	60	40	0	0	1	1.3	60	19.0
142	0	5	0	95	0	80	20	0	0	1	3.8	0	19.0
143	0	10	0	90	0	70	30	0	0	1	6.3	0	19.5
144	0	50	0	0	0	15	30	0	0	0	0.0	100	21.3
145	0	20	0	80	0	50	50	0	0	1	2.8	0	21.3
146	0	0	0	100	0	90	10	0	0	0	5.3	0	22.0
147	0	0	0	100	0	75	25	0	0	0	7.5	0	22.0
148	0	90	0	10	0	60	30	10	0	1	1.3	20	23.8
149	0	0	0	100	0	80	10	10	0	0	3.8	0	23.8
150	0	0	0	100	0	60	40	0	0	0	6.3	0	23.8
151	0	0	40	60	40	50	50	0	0	1	8.3	0	23.9

*Variables are: X_1 = percentage of soils with constant lime enrichment; X_2 = percentage meadow soil with calcium groundwater; X_3 = percentage soils formed from coral bedrock under conditions of constant lime enrichment; X_4 = percentage alluvial and organic soils adjacent to rivers and saline organic soil at the coast; X_5 = percentage deciduous seasonal broadleaf forest; X_6 = percentage high and low marsh forest, herbaceous marsh and swamp; X_7 = percentage cohune palm forest; X_8 = percentage mixed forest composed of types listed for X_5 and X_7; X_9 = number of soil boundaries in square; X_{10} = distance to navigable water (km); X_{11} = percentage of square within 1 km of navigable water; X_{12} = distance from the site of Santa Rita (km).

6 Estimating a resource selection probability function from a census of resource units at several points in time using the proportional hazards model

When used and unused resource units are censused after several periods of selection time it is necessary for an analysis to take into account the increasing proportion of used units as time increases. One way to do this involves modelling data using the proportional hazards survival model as discussed in this chapter.

6.1 INTRODUCTION

In the last chapter logistic regression was suggested as a method for estimating a resource selection probability function for census data when there is no need to take into account variable amounts of selection time. In the present chapter an alternative model that takes into account the selection time is proposed in order to analyse data involving two or more periods of selection, possibly of different durations.

The situation envisaged is that a population of A_+ available resource units is subjected to S selection episodes following one after another, with a census of resource units at the end of each episode. The structure of the selection process is then as indicated in Figure 6.1. Thus the available resource units at time 0 are divided at time t_1 into a group of unused units and a group of used units. Then, during the second selection episode, from time t_1 to time t_2, some of the unused units at time t_1 become used. The same process continues until time t_S and, in general, as the selection time increases there are fewer and fewer unused units, and more and more used units. Note that this description of a selection process holds for resource units that can be used more than once since it is the first use that changes the status of a unit.

As in the last chapter, it is assumed that each resource unit is characterized by the values that it possesses for p variables $X_1, X_2,..., X_p$, and that the A_+ re-

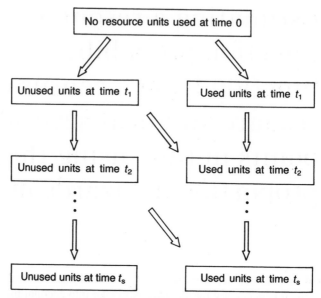

Figure 6.1   The structure of the resource selection process when resource units can be used only once and censuses of units are made after selection times of $t_1, t_2, ..., t_S$.

source units initially available can be divided into I classes so that all the A_i units within the ith class have the X values $x_i = (x_{i1}, x_{i2}, ..., x_{ip})$. The question at issue then becomes how the probability of use by time t for units in the ith class depends on x_i.

From censuses of resource units taken at times t_1 to t_S, and a knowledge of what resource units were available at time 0, it will be possible to determine u_{ij}, the number of type i units used between times t_{j-1} and time t_j, and $\bar{u}_i$, the number of type i units that are unused at time t_S. Hence the data resulting from the study can be set out in the form indicated in Table 6.1.

Table 6.1   A format for the data arising from taking censuses of resource units on S occasions

Type of unit	X values				Initial number	Used during the time period				Unused at time t_S
	X_1	X_2	...	X_p		$0-t_1$	t_1-t_2	...	$t_{S-1}-t_S$	
1	x_{11}	x_{12}	...	x_{1p}	A_1	u_{11}	u_{12}	...	u_{1S}	$\bar{u}_1$
2	x_{21}	x_{22}	...	x_{2p}	A_2	u_{21}	u_{22}	...	u_{2S}	$\bar{u}_2$
.										
.										
.										
I	x_{I1}	x_{I2}	...	x_{Ip}	A_I	u_{I1}	u_{I2}	...	u_{IS}	$\bar{u}_I$

Clearly, the selection of resource units is a type of survival process so that it is appropriate to model the resource selection probability function by

$$w^*(x,t) = 1 - \phi^*(x,t), \tag{6.1}$$

where $\phi^*(x,t)$ is a standard survival function. There are many possible choices for $\phi^*(x,t)$, but the only one that will be considered in this chapter is

$$\phi^*(x,t) = \exp\{-\exp(\beta_0+\beta_1 x_1+...+\beta_p x_p)t\}. \tag{6.2}$$

This is sometimes called the proportional hazards survival function since, for all selection times t, the hazard function for type i resource units, which is $\exp(\beta_0+\beta_1 x_{i1}+...+\beta_p x_{ip})t$, is proportional to the hazard function for type k resource units, which is $\exp(\beta_0+\beta_1 x_{k1}+...+\beta_p x_{kp})t$. The use of equation (6.2) is realistic in the present context since it implies that $w^*(x,0) = 0$ (no resource units are used when t=0) and $w^*(x, \infty) = 1$ (all resource units are eventually used).

If each resource unit is used or not used independently of other units then the distribution of the numbers of type i resource units used in the intervals $(0, t_1)$, $(t_1,t_2),..., (t_{s-1},t_s)$, and the number of this type of unit that are unused at time t_s, will follow a multinomial distribution such that the probability of use in the time interval (t_{j-1},t_j) is

$$\begin{aligned}
\Theta_{ij} &= \phi^*(x_i,t_{j-1}) - \phi^*(x_i,t_j) \\
&= \exp\{-\exp(\beta_0+\beta_1 x_{i1}+...+\beta_p x_{ip})t_{j-1}\} \\
&\quad - \exp\{-\exp(\beta_0+\beta_1 x_{i1}+...+\beta_p x_{ip})t_j\},
\end{aligned} \tag{6.3}$$

and the probability of not being used by time t_s is

$$\Theta_{i\,s+1} = \exp\{-\exp(\beta_0+\beta_1 x_{i1}+...+\beta_p x_{ip})t_s\}. \tag{6.4}$$

6.2 ESTIMATING THE PROPORTIONAL HAZARDS FUNCTION

The model defined by equations (6.1) to (6.4) can be fitted to data of the form shown in Table 6.1 using the principle of maximum likelihood. All that is required is a computer program for fitting multinomial data such as MAXLIK (Reed and Schull, 1968; Reed, 1969, Manly, 1985, p. 433), or GLIM (McCullagh and Nelder, 1989).

One of the limitations of using the survival function of equation (6.2) is that there is the implicit assumption that the rate at which units are used remains more or less constant. Hence, for example, the model of equations (6.1) to (6.4) may not fit a set of data because the proportion of resource units used is very large or very small for a particular time interval between two censuses. This can occur, although the relative preferences for different types of resource unit remain constant.

One way to allow for this type of effect involves recognizing that it amounts to the same thing as the effective selection time being different from the chronological time. Thus selection time runs slowly when the rate of selection of units is low, and runs quickly when the rate of selection of units is high. This can be allowed for by setting t_1 equal to the nominal selection time but regarding t_2 to t_s

as unknown 'effective' selection times to be estimated along with the β parameters. It is not possible to estimate t_1 along with the other selection times since if that is done then β_0 in equations (6.2) to (6.4) becomes confounded with all the selection times. Fixing t_1 is necessary in order to determine the scale for measuring the other times.

As discussed in section 3.3, minus twice the maximized log-likelihood will follow a distribution that is approximately chi-squared if the model being fitted is correct, with the degrees of freedom being the number of data frequencies minus the number of estimated parameters. In the present context, the statistic found in this way is

$$X_L^2 = 2[\sum_{i=1}^{I} \sum_{j=1}^{S} u_{kj}\log_e\{u_{ij}/E(u_{ij})\} + \sum_{i=1}^{I}\bar{u}_i\log_e\{\bar{u}_i/E(\bar{u}_i)\}], \qquad (6.5)$$

with $I(S+1)-p-1$ degrees of freedom, where $E(u_{ij})$ and $E(\bar{u}_i)$ denote the expected values from the fitted model for counts of used and unused type i resource. This statistic can be used for a goodness-of-fit test providing that the expected frequencies are 'large', and differences between X_L^2 values can be used to compare the fit of different models even if this is not the case.

If the assumption of a multinomial distribution is correct for the counts u_{i1}, $u_{i2},\ldots,$ u_{iS} and $\bar{u}_i$ of used and unused type i resource units then the individual counts will follow binomial distributions. Hence the mean and variance of u_{ij} will be $A_i\Theta_{ij}$ and $A_{ij}\Theta_{ij}(1-\Theta_{ij})$, respectively, and the mean and variance of $\bar{u}_i$ will be $A_i\Theta_{iS+1}$ and $A_{ij}\Theta_{iS+1}(1-\Theta_{iS+1})$, respectively, where the Θ values are defined by equations (6.3) and (6.4). Thus standardized residuals to show the discrepancy between observed and expected counts can be defined as differences divided by their estimated standard deviations, which are

$$R_{ij} = (u_{ij} - A_i\hat{\Theta}_{ij})/\sqrt{\{A_i\hat{\Theta}_{ij}(1 - \hat{\Theta}_{ij})\}}, \qquad (6.6)$$

and

$$R_{iS+1} = (\bar{u}_i - A_i\hat{\Theta}_{iS+1})/\sqrt{\{A_{ij}\hat{\Theta}_{iS+1}(1-\hat{\Theta}_{iS+1})\}}, \qquad (6.7)$$

where 'caps' over the Θs indicate the values obtained from equations (6.3) and (6.4) using estimated β parameters. As usual, it is desirable that most of the standardized residuals should be within the range from -2 to $+2$, and almost all within the range from -3 to $+3$. In addition, the residuals should show no patterns when plotted against expected frequencies, values of X variables, time, etc.

6.2.1 Example 6.1 Selection of snails by birds

As an example, consider the experiment of Bantock *et al.* (1976) on the selection of *Cepaea nemoralis* and *C. hortensis* snails by the song thrush *Turdus ericetorum*, that was described in example 2.3. Recall that an experimental population of 498 yellow five-banded (Y5H) *C. hortensis*, 499 yellow five-banded (Y5N) *C. nemoralis* and 877 yellow mid-banded (Y3N) *C. nemoralis* snails was set up on 29 June, 1973, with the shells being uniquely marked so that the survivors could be determined from censuses taken at various times after the population was set up. Extra Y3H were added to the population on 5 July and on 7 July. This complicates any analysis that includes this morph and therefore for the purposes of this example this morph will be ignored and only the results for used and unused Y5N and Y5H will be considered.

Simplified results from the experiment are shown in Table 2.3, with different types of snail defined in terms of two variables, X_1, a species indicator which is 1 for *C. nemoralis* and 0 for *C. hortensis*, and X_2 the maximum shell diameter in units of 0.3 mm over 14.3 mm.

Various versions of the proportional hazards model defined by equations (6.1) to (6.4) have been fitted to the data using the computer program RSF that is described in the Preface. First, the 'no selection' model, with only the β_0 term in equations (6.2) to (6.4) was fitted using the known survival times of 6, 12 and 22 days. This model, which will be referred to as model 0A, gave a log-likelihood chi-squared goodness-of-fit statistic of 205.81 with 125 degrees of freedom. The 'no selection' model with the last two selection times estimated was also fitted. This model, which will be called model 0B, gave a goodness-of-fit log-likelihood chi-squared value of 196.31 with 123 degrees of freedom. The difference in chi-squared values between model 0A and model 0B is therefore 9.50 with 2 degrees of freedom. Since this is significantly large at the 1% level, it seems that model 0B is more satisfactory than model 0A.

Next, models allowing for selection related to species and size were fitted by including X_1 and X_2 in equations (6.2) to (6.4). As for the 'no selection' case, two models were considered. For model 1A the chronological times of 12 and 22 days were used for t_2 and t_3. This then resulted in a log-likelihood goodness-of-fit statistic of 143.19, with 123 degrees of freedom. For model 1B the second and third sample times were treated as parameters to be estimated, with the result that the goodness-of-fit statistic dropped to 133.45, with 121 degrees of freedom. The difference in the goodness-of-fit statistics for model 1A and model 1B is 9.74, with two degrees of freedom, which is significantly large at the 1% level. Hence the model with estimated sample times is a distinctly better fit than the model using chronological times.

Since many of the expected sample frequencies are very small, it is questionable whether it is valid to compare the model 1B goodness-of-fit statistic of 133.45 with 121 degrees of freedom with the chi-squared distribution as a test for the absolute goodness-of-fit. However, if this is done then the statistic is found to be not at all significantly large.

Models involving effects that are quadratic in time were also fitted by including $X_3 = X_2^2$ in equations (6.2) to (6.4). Since the addition of this extra variable led to very little reduction in goodness-of-fit statistics, these models will not be considered further here. Models that allowed the coefficient of the size variable to vary with the species were also fitted, but again this led to little reduction in goodness-of-fit statistics so these will not be discussed further here.

Table 6.2   Estimated parameters for the proportional hazards equations (6.2) to (6.4)

Parameter	Estimate	Standard error	Ratio
Constant, β_0	−3.55	0.27	−12.99
Size effect, β_1	0.024	0.014	1.80
Species effect, β_2	−0.48	0.13	−3.60
Effective selection time, t_2	14.07	0.83	16.88
Effective selection time, t_3	24.40	1.67	14.58

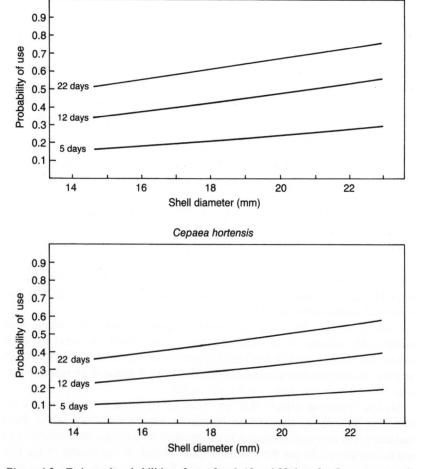

Figure 6.2   Estimated probabilities of use after 6, 12 and 22 days for *Cepaea nemoralis* and *C. hortensis*.

It seems from this analysis that model 1B is realistic for the data. The estimated parameters found for this model are shown in Table 6.2, and it can be seen by considering the ratios of β estimates to their standard errors that there is clear evidence of a species effect but little evidence of a size effect. It can also be seen that the estimated selection times t_2 and t_3 are not too different from the chronological times of 12 and 22 days.

The resource selection probability function for model 1B is given by Table 6.2 to be

$$w^*(x,t) = 1 - \exp[-\exp\{-3.55 + 0.024(SIZE) - 0.48(SPECIES)\}t],$$

where SPECIES denotes the dummy variable for the species (0 for *C. hortensis*, 1 for *C. nemoralis*) and SIZE denotes the coded size of snails. Because the second and third selection times were estimated, probabilities can only be determined from this function for the census times of six days (using $t = 6$), 12 days (using $t = 14.07$) and 22 days (using $t = 24.40$).

Figure 6.2 shows the level of selection that is suggested by model 1B. There is apparently a rather higher probability of use for *C. nemoralis* than for *C. hortensis*, and for large snails rather than small ones. However, it must be kept in mind that the size effect is not significant at the 5% level. Figure 6.3 shows standardized residuals calculated using equations (6.6) and (6.7). A separate graph is provided for the residuals based on the numbers used up to day 6, the numbers used between day 6 and day 12, the number used between day 12 and day 22, and the number unused on day 22. The species involved with a residual is indicated by 'H' or 'N'. All the plots indicate the type of distribution expected from standard normal variables so that it seems that the model gives an adequate description of the data.

6.3 DISCUSSION

The proportional hazards model that is described in this chapter is appropriate for analysing data from studies where full censuses of resource units are taken after several periods of selection time. The proportional hazards model is only one of many that might be considered in this type of situation (McCullagh and Nelder, 1989, Chapter 13), but in fact the methods suggested here are general in the sense that they can be used equally well with any other survival function used for equation (6.2) that is a function of the X variables measured on resource units.

Computer programs for fitting the proportional hazards model to survival data are fairly readily available, and can be used to carry out the calculations that have been described in this chapter. However, researchers may find it more convenient to use the program RSF that is described in the Preface since this is a program that is designed specifically for use in the context of resource selection.

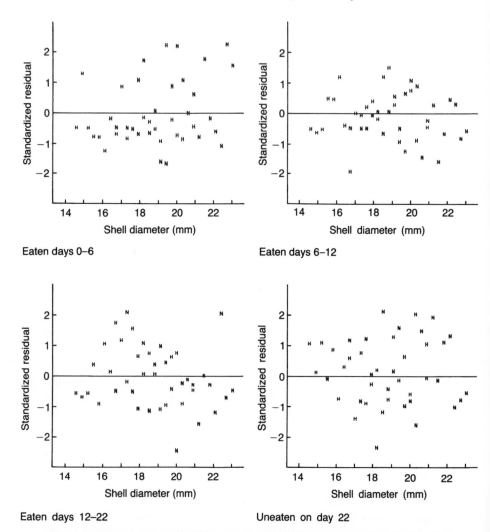

Figure 6.3 Standardized residuals plotted against the maximum shell diameters of snails. A separate graph is given for the numbers eaten in each of the time intervals 0–6, 6–12 and 12–22 days, and the numbers uneaten on day 22 (H is for *Cepaea hortensis*, N for *C. nemoralis*).

Exercise

Example 5.1 was concerned with Ryder's (1983) study of winter habitat selection by pronghorn (*Antilocapra americana*) in the Red Rim area of south-central Wyoming. This study involved recording a number of variables on each of 256 plots of ground in the region, and noting whether pronghorn were seen on each plot in the winters of 1980–81 and 1981–82. From the discussion in example 5.1 it seems that the probability of a plot being used was mainly a function of the distance to water and the aspect of the plots.

For the present exercise, assume that the distance to water and the aspect of a plot are the only important variables for determining probability of use. On this basis, it is possible to group the 256 study plots into I = 36 different types of unit and hence fit the proportional hazards model to the data providing that a plot is considered to be used the first time that pronghorn are recorded. In this way, the extensive set of data provided in Table 2.2 reduces to the much smaller set shown in Table 6.3.

Table 6.3 The data on winter habitat use in Table 2.2 arranged as required for the estimation of the model of equations (6.1) to (6.4) when only the distance to water and the aspect of plots are considered

Type of study plot	Distance to water $X_1(m)$	Dummy variables for aspect			u_1	u_2	$\bar{u}$
		X_2	X_3	X_4			
1	25	1	0	0	1	1	0
2	150	1	0	0	2	0	0
3	375	1	0	0	5	0	0
4	625	1	0	0	1	0	2
5	875	1	0	0	0	2	1
6	1250	1	0	0	1	0	3
7	1750	1	0	0	0	0	1
8	2250	1	0	0	1	0	2
9	2750	1	0	0	2	2	1
10	25	0	1	0	0	0	1
11	150	0	1	0	0	2	0
12	375	0	1	0	1	0	1
13	625	0	1	0	0	0	2
14	875	0	1	0	2	0	0
15	1250	0	1	0	0	0	4
16	1750	0	1	0	2	0	3
17	2250	0	1	0	1	1	1
18	2750	0	1	0	0	1	9
19	25	0	0	1	0	0	1
20	150	0	0	1	1	1	0
21	375	0	0	1	4	1	0
22	625	0	0	1	3	0	6
23	875	0	0	1	6	1	3
24	1250	0	0	1	0	1	4
25	1750	0	0	1	0	0	4
26	2250	0	0	1	1	2	1
27	2750	0	0	1	1	1	6
28	25	0	0	0	4	4	5
29	150	0	0	0	2	2	5
30	375	0	0	0	5	4	8

Table 6.3   (Cont'd)

Type of study plot	Distance to water X_1 (m)	Dummy variables for aspect			u_1	u_2	$\bar{u}$
		X_2	X_3	X_4			
31	625	0	0	0	6	2	10
32	875	0	0	0	6	11	10
33	1250	0	0	0	2	3	15
34	1750	0	0	0	6	6	6
35	2250	0	0	0	1	0	5
36	2750	0	0	0	5	6	13

The three aspect variables are $X_2 = 1$ for East/Northeast, otherwise 0, $X_3 = 1$ for South/Southeast, otherwise 0, and $X_4 = 1$ for West/Southwest, otherwise 0. The fourth aspect, North/Northwest, is the 'standard' aspect that receives 0 for X_2 to X_4. The use variables are u_1 = number used in winter 1, u_2 = number used for the first time in winter 2, and $\bar{u}$ = number not used.

Fit the model of equations (6.1) to (6.4) to this reduced set of data using the principle of maximum likelihood, and compare the estimated resource selection probability function that is obtained in this way with the function that was estimated by logistic regression in example 5.1. Examine standardized residuals to see whether the model appears to give a reasonable fit to the data for all aspects and distances to water.

7 Estimating resource selection functions from samples of resource units using proportional hazards and log-linear models

In this chapter, three models for sample resource selection data are discussed, where the samples can be of available resource units, used resource units or unused resource units. The differences between these models are related to the sampling protocols A, B and C that have been defined in section 1.4, but an extra important factor is the proportion of available units that are used. The first model is appropriate when either sampling fractions are known or a large proportion of resource units are used, and samples of used and unused units are taken. The second model is appropriate when only a small fraction of the available resource units are used and samples of available and used resource units are available. The third model is appropriate when an appreciable proportion of resource units are used and only samples of unused units are taken.

7.1 INTRODUCTION

The situation considered in the previous chapter was where a population of N available resource units is censused at time 0, before any are selected, and then at subsequent times $t_1, t_2,..., t_S$, as indicated in figure 6.1. It was also assumed that the N units could be divided into I groups of resource units such that within the ith group all units have the same values $x_i=(x_{i1},x_{i2},...,x_{ip})$ for certain characterizing variables X_1 to X_p. On this basis a proportional hazards model was proposed for the resource selection probability function $w^*(x)$.

In the present chapter exactly the same type of selection process is considered, with more and more units being used as time passes. However, instead of assuming that censuses of units take place at times 0, $t_1, t_2,..., t_S$, it will be assumed that random samples of used or unused units are taken at some or all of these times.

There are a number of study designs within this framework that are commonly used, with perhaps the most popular involving the collection of a sample of available units and an independent sample of used units after one period of selection. The study of nest selection by fernbirds (*Bowdleria puncta*) in Otago, New Zealand, that was the subject of example 2.4 is a typical example of this type. It may be recalled that Harris (1986) measured several variables on 24 nest sites found during the 1982–83 and 1983–84 seasons, and found comparative available sites by choosing 25 random points in the study area.

Although most studies consider only a single selection period, there are cases where a population of resource units has been sampled at several times while selection is proceeding. One such case was discussed in example 2.5, which was concerned with Popham's (1944) study of the use of corixids as food by minnows (*Phoxinus phoxinus*). Here, Popham sampled the corixid in a pond every day for seven days, introduced 50 minnows to the pond on the evening of the seventh day, and then sampled again every day from the third to ninth days after this change to the pond. The samples taken before the introduction of minnows can be lumped together to form a single 'available sample', since they have similar proportions for the species and colours of corixids, while the samples taken after the introduction of minnows are samples of unused prey. Therefore this can be thought of as a situation where there are eight samples of unused resource units (corixids), taken after selection times of 0, 3, 4,..., 9 days.

7.2 EXPECTED SAMPLE FREQUENCIES FOR RESOURCE UNITS

In this section equations for expected sample frequencies of resource units are developed based on the assumptions stated in the last section that all the sampled units fall within one of I classes of units such that all the units within class i have the vector of values x_i for the variables X_1 to X_p. The motivation for developing these equations comes from the fact that they depend on the resource selection probability function and can therefore be used as a means of estimating this function. More will be said about estimation in section 7.6. At this point it will merely be noted that procedures exist whereby maximum likelihood estimates of unknown parameters can be calculated using equations for expected sample frequencies, providing that it is reasonable to assume that the observed sample frequencies have Poisson distributions with these expected values.

There are three types of samples that need to be considered: (a) available resource units before selection begins, (b) used units at time t, and (c) unused units at time t. To find expected sample frequencies for the first type of sample, suppose that at time 0, before selection begins, there are A_i available units in the ith class, of which a_i are seen in a random sample. Then the expected value of a_i is

$$E(a_i) = P_a . A_i,$$ (7.1)

where P_a is the sampling probability. In developing this equation and the equations that follow it is assumed that when a sample is taken each resource unit in the relevant universe has the same probability of being recorded, independent of what happens for any other unit.

Sometimes the sampling scheme may be different in the sense that the total sample size is fixed in advance. However, in that case the equations will still hold but with the sampling probabilities being redefined to be sampling fractions. Hence from now on in this chapter it should be understood that the expression 'sampling probability' can also mean 'sampling fraction'.

As the selection time increases, more and more of the unused units in the ith class will become used units, until eventually no more unused units are left. To allow for this, it will be assumed that if a sample of used units is taken at time $t > 0$ then the expected value of $u_i(t)$, the number of type i individuals in the sample, is

$$E\{u_i(t)\} = P_u(t).A_i.w^*(x_i,t), \tag{7.2}$$

where $P_u(t)$ is the sampling probability for used units at time t, and $w^*(x_i,t)$ is the resource selection probability function which gives the probability that a type i unit will be used by time t. Equation (7.2) still applies in situations where units can be used more than once if 'use' is defined to be use at least once. Also, since no units are used at time $t = 0$, it follows that $w^*(x_i,0) = 0$, and hence that $E\{u_i(0)\} = 0$.

An equation can also be developed for the expected value of $\bar{u}_i(t)$, the number of class i individuals in a sample of unused units that is taken at time t. This is

$$E\{\bar{u}_i(t)\} = P_{\bar{u}}(t).A_i.\{1 - w^*(x_i,t)\}, \tag{7.3}$$

where $P_{\bar{u}}(t)$ is the sampling probability for unused units at time t. Since the unused units at time 0 are the available units, equation (7.3) with $t = 0$ is essentially the same as equation (7.1). Formally this identification is achieved by recognizing that the following relationships must hold: $\bar{u}_i(0) = a_i$, $P_a = P_{\bar{u}}(0)$, and $w^*(x_i,0) = 0$.

7.3 THE PROPORTIONAL HAZARDS MODEL

At this point the function $w^*(x_i,t)$ in equations (7.2) and (7.3) can take any form as long as the condition $0 \le w^*(x_i,t) \le 1$ is satisfied. However, in order to estimate the function it is necessary to assume a specific parametric form. To be consistent with the previous chapter, it will be assumed that the probability of *not* being used by time t is given by the proportional hazards function of equation (6.2) so that

$$w^*(x_i,t) = 1 - \exp\{-\exp(\beta_0 + \beta_1 x_{i1} + ... + \beta_p x_{ip})t\}. \tag{7.4}$$

Equations (7.1) to (7.4) between them describe the expected values of sample frequencies in terms of the parameters A_1 to A_I, P_a, $P_u(t)$, $P_{\bar{u}}(t)$, and β_0 to β_p. In principle it is therefore possible to estimate some or all of these parameters if observed sample frequencies are available for two or more samples either of different types of units or taken at different times. There are, however, some difficulties involved in estimating all the parameters because of the way that they combine together.

Two special cases that merit further discussion in this connection occur when the proportion of used units is very small, or when only samples of unused units are available. These two cases will now be considered in turn.

7.4 SMALL PROPORTION OF RESOURCE UNITS USED

The case where only a small proportion of available resource units are used during the study period occurs often. This is, for example, the situation with the study on the selection of nest sites mentioned above since only a very small proportion of potential nest sites were used. What happens under these circumstances is that the exponential term $\exp(\beta_0 + \beta_1 x_{i1} + ... + \beta_p x_{ip})t$ is small in equation (7.4), so that

$$
\begin{aligned}
w^*(x_i,t) &= 1 - \exp\{-\exp(\beta_0 + \beta_1 x_{i1} + ... + \beta_p x_{ip})t\} \\
&\approx 1 - \{1 - \exp(\beta_0 + \beta_1 x_{i1} + ... + \beta_p x_{ip})t\} \\
&\approx \exp(\beta_0 + \beta_1 x_{i1} + ... + \beta_p x_{ip})t.
\end{aligned}
\tag{7.5}
$$

Also, the difference between the universe of available resource units and the universe of unused units is negligible. Hence equations (7.2) and (7.3) become

$$
E\{u_i(t)\} = P_u(t).A_i.\exp(\beta_0 + \beta_1 x_{i1} + ... + \beta_p x_{ip})t
\tag{7.6}
$$

and

$$
E\{\bar{u}_i(t)\} = P_{\bar{u}}(t).A_i.
\tag{7.7}
$$

Equation (7.6) can be rewritten as

$$
E\{u_i(t)\} = P_u(t).A_i.\exp(\beta_0).\exp(\beta_1 x_{i1} + ... + \beta_p x_{ip})t,
$$

which shows that the parameter β_0 in the resource selection probability function is confounded with the sampling probabilities $P_u(t)$ in the sense that any change to the value of β_0, and hence $\exp(\beta_0)$, can be adjusted for completely by appropriate changes in the $P_u(t)$ parameters. What this means in practice is that the product $P_u(t)\exp(\beta_0)$ can be estimated, but not both $P_u(t)$ and β_0.

This is not a problem when sampling probabilities are known since then β_0 can be estimated. When both the parameters are unknown it is possible to produce a model that can be estimated by setting $\beta_0 = 0$ in equation (7.6) and recognizing that in effect the apparent estimate $P_u(t)$ is really an estimate of $P_u(t)\exp(\beta_0)$. It is then still possible to estimate

$$
w(x_i,t) = \exp(\beta_1 x_{i1} + \beta_2 x_{i2}... + \beta_p x_{ip})t,
\tag{7.8}
$$

which is a resource selection function since it is proportional to the resource selection probability function. As will be shown in the examples that follow, this resource selection function is almost as useful as the resource selection probability function itself, so that not being able to estimate β_0 is no real hardship.

7.5 ONLY SAMPLES OF UNUSED RESOURCE UNITS AVAILABLE

Suppose that only samples of unused resource units are taken and sampling probabilities are unknown. In that case it is convenient to approximate the resource selection probability function by

$$w^*(x_i,t) = 1 - \exp\{(\beta_0 + \beta_1 x_{i1} + \ldots + \beta_p x_{ip})t\}, \qquad (7.9)$$

where the argument of the exponential function should be negative, so that $w^*(x_i,0) = 0$ and $w^*(x_i,\infty) = 1$.

Substitution of function (7.9) into equation (7.3) yields

$$\begin{aligned}E\{\bar{u}_i(t)\} &= P_{\bar{u}}(t).A_i.\exp\{(\beta_0 + \beta_1 x_{i1} + \ldots + \beta_p x_{ip})t\}, \\ &= P_{\bar{u}}(t).A_i.\exp(\beta_0 t).\exp\{(\beta_1 x_{i1} + \ldots + \beta_p x_{ip})t\}.\end{aligned} \qquad (7.10)$$

The last equation shows that β_0 is confounded with $P_{\bar{u}}$ since a change in β_0, and hence $\exp(\beta_0 t)$, can be compensated for by changing the value of $P_{\bar{u}}(t)$ in an appropriate way. Hence, if $P_{\bar{u}}(t)$ is unknown then it is not possible to estimate β_0. The obvious way round this difficulty is to arbitrarily set $\beta_0 = 0$, and recognize that this implies that the apparent estimate of $P_{\bar{u}}(t)$ is really an estimate of $P_{\bar{u}}(t)\exp(\beta_0 t)$. The function

$$\phi(x_i,t) = \exp\{(\beta_1 x_{i1} + \beta_2 x_{i2} + \ldots + \beta_p x_{ip})t\}, \qquad (7.11)$$

which can be estimated, then gives the probability that a unit is unused at time t, multiplied by an unknown constant.

From an estimate of $\phi(x_i,t)$ it is possible to order resource units on the basis of their estimated probabilities of not being used by time t, which will be the opposite to the order for their probability of being used. This is not as satisfactory as estimating the resource selection probability function, but may give sufficient information to make a study worthwhile.

It is recommended that equation (7.9) be used instead of equation (7.4) to describe data frequencies whenever samples of used units are not available and sampling probabilities are unknown. The reason for making this recommendation is that the confounding that is present between the resource selection probability function and the sampling probabilities is largely removed when β_0 is set to zero. If an attempt is made to estimate the full resource selection probability function under these circumstances then the confounding will still be present, although this is not so transparent. The net result will be that estimates of all the parameters in the model will be adversely affected and it will not be possible to determine any of them well.

7.6 ESTIMATING PARAMETERS FROM SAMPLE DATA

The equations for expected sample frequencies that are provided in the last three sections can serve as a basis for estimating the parameters of a resource selection probability function, providing that at least two samples of resource units are taken, and these samples differ because they are of a different nature (such as used and unused units), or differ in the time when they are taken.

The equations for expected sample frequencies are not by themselves sufficient to estimate the parameters in resource selection probability functions by the method of maximum likelihood. It is also necessary to make assumptions about the distributions of sample frequencies. The simplest possibility here involves assuming that sample counts a_i, $u_i(t)$ and $\bar{u}_i(t)$ have independent Poisson distributions, with expected values given by the equations in the previous sections. This is reasonable if samples are taken independently, are random, and sample sizes are much smaller than the universes being sampled.

Given the Poisson assumption, estimates of unknown parameters can be obtained using one of the available general computer programs for the maximum likelihood estimation of models of this form. For the examples presented below, the program RSF that is described in the Preface has been used. This program gradually improves initial approximations for maximum likelihood estimates of unknown parameters, and produces large sample variances and covariances for the final estimates, as well as chi-squared goodness-of-fit statistics.

There will be one A_i parameter for each of the I distinct types of resource unit so that there may be a large number of these parameters. This implies that the iterative process that is required by computer programs for maximum likelihood estimation may become very slow since each iteration involves inverting a matrix where the numbers of rows and columns is equal to the total number of parameters to be estimated. However, there is a 'trick' that can be used to overcome this problem since the maximum likelihood estimate of A_i is the value that makes the total expected number of units of the ith type equal to the total number observed (Manly, 1985, p. 435). What has to be done is to estimate the sampling probability and resource selection function parameters by iteration, and at the end of each iteration estimate A_i by equating the total observed and expected frequency of type i units. It can be shown that if the iterations converge then this process will produce maximum likelihood estimates of all the parameters, and the correct large sample estimated variances and covariances for the sampling probability and resource selection function parameters.

In discussing the details of estimation it is necessary to consider three different models, corresponding to the material in sections 7.3, 7.4 and 7.5. For convenience it is useful to refer to these from now on as the 'general' model, the 'small fraction used' model, and the 'unused units only' model. These will now be considered in turn.

7.6.1 The general model

Equations (7.1) to (7.3) involve $S+I+p+1$ parameters when the resource selection probability function of equation (7.4) is substituted for $w^*(x_i,t)$: P_a,

$P_u(t)$ and $P_{\bar{u}}(t)$ for S sample times, A_1 to A_I, and β_0 to β_p. It is always helpful if the sampling probabilities are known since if they are not then at best it will only be possible to estimate their relative values. This is because of the way that the sampling probabilities always appear as multipliers of the A_i parameters, which means that if all the sampling probabilities are multiplied by a positive constant then it is possible to compensate for this exactly in equations (7.1) to (7.3) by dividing the A_i values by the same constant.

This inherent confounding of parameters does not affect the estimation of the parameters of the resource selection probability function and can be overcome easily by arbitrarily setting one of the sampling probabilities equal to 1. In effect, the estimates of the other sampling 'probabilities' then become relative to this fixed parameter. In a way the confounding is even useful since it means that any known measures of sampling effort that are proportional to sampling probabilities can be substituted for the sampling probabilities in equations (7.1) to (7.3) when estimating the model. The constant multiplying factor that changes the measures of sampling effort into probabilities will then automatically be absorbed into the A_i parameters.

As explained above, the maximum likelihood estimates of the A_i parameters can be calculated as functions of the other parameters with a potentially considerable saving of the time required for calculations.

7.6.2 The small fraction used model

Section 7.4 was concerned with the special case when only a small fraction of the resource units are used during the study period and sampling fractions are unknown. Under these circumstances the parameter β_0 in the resource selection probability function cannot be estimated. However, the resource selection function given by equation (7.8) can be estimated by setting β_0 equal to zero, with sample frequencies given by equations (7.6) and (7.7), providing that there is at least one sample of used units and at least one sample of either unused or available units.

As with the estimation for the general model, there are three types of parameter involved in equations (7.6) and (7.7): the sampling probabilities $P_u(t)$ and $P_{\bar{u}}(t)$, the parameters A_1 to A_I for the available frequencies of different types of unit, and the parameters β_1 to β_p of the resource selection function. Also, as was the situation with the general model, the sampling probabilities multiply the A_i parameters. Hence it is not possible to estimate the sampling probabilities and the A_i parameters. However, if one of the sampling probability parameters is set equal to 1 then the other parameters of this type become relative sampling probabilities, which can be estimated.

One of the useful aspects of estimation in situations where the small fraction used model applies is that the sample frequencies of different types of resource unit have log-linear models. This is useful since it means that many of the available computer programs for log-linear modelling can be used to estimate parameters.

To see how a log-linear model formulation applies, consider first equation (7.6) for the expected frequency of type i resource units in a sample of used units taken at time t. Setting $\beta_0 = 0$, this equation can be written in the form

$$E\{u_i(t)\} = \exp[\log_e\{tP_u(t)\}+\log_e(A_i)+\beta_1 x_{i1}+ \ldots +\beta_p x_{ip}],$$

so that the logarithm of the expected frequency is given by a linear combination of the parameters $\log_e\{tP_u(t)\}$, $\log_e(A_i)$, and β_1 to β_p. The assumption that $u_i(t)$ follows a Poisson distribution then makes this a conventional log-linear model. In a similar way, equation (7.7) for the expected frequency of type i resource units in a sample of unused units can be written as

$$E\{\bar{u}_i(t)\} = \exp[\log_e\{P_{\bar{u}}(t)\}+\log_e(A_i)],$$

so that the logarithm of the expected frequency is a linear combination of the parameters $\log_e\{P_{\bar{u}}(t)\}$ and $\log_e(A_i)$. Again, the assumption that $\bar{u}_i(t)$ follows a Poisson distribution makes this a conventional log-linear model.

7.6.3 The unused units only model

In section 7.5 the situation has been discussed where sampling probabilities are unknown, only samples of unused units are available, and the number of units used during the study is sufficient to change the proportions of different types of resource units by appreciable amounts. Given this situation, the resource selection probability function can be approximated by equation (7.9), so that expected sample frequencies of different types of resource unit are given by equation (7.10). To identify parameters, β_0 must be set equal to 0 in equation (7.10). The probability of a unit not being used by time t is then proportional to the function $\phi(x_i,t)$ of equation (7.11).

As with the general model and the small fraction used model, there are three types of parameter involved in equation (7.10). With S samples, these are the sampling probabilities $P_{\bar{u}}(t_1)$ to $P_{\bar{u}}(t_S)$, the parameters for the available frequencies of different types of unit A_1 to A_I, and the parameters β_1 to β_p of the resource selection probability function. To overcome confounding between the first two types of parameter, $P_{\bar{u}}(t_1)$ can be set equal to 1, so that $P_{\bar{u}}(t_2)$ to $P_{\bar{u}}(t_S)$ become sampling probabilities relative to this.

Equation (7.10) with $\beta_0 = 0$ can be rewritten as

$$E\{\bar{u}_i(t)\} = \exp[\log_e\{P_{\bar{u}}(t)\}+\log_e(A_i)+(\beta_1 x_{i1}+ \ldots +\beta_p x_{ip})t].$$

This is then a log-linear model on the assumption that $\bar{u}_i(t)$ follows a Poisson distribution. Thus the parameters for the unused units only model can be estimated by a program for log-linear modelling if this is desirable.

7.7 INFERENCES WITH SAMPLE DATA

The modelling procedures that have been described in Chapter 3 can be used with sample data. Samples that differ either in the time that they were collected or the nature of the resource units (available, used or unused) can be compared using standard statistical methods in order to get a general understanding of the nature of sample differences. A check can be made to see whether the estimate $\hat{\beta}_i$ of the parameter β_i in the resource selection probability function is significantly different from zero by comparing $\hat{\beta}_i/\text{se}(\hat{\beta}_i)$ with critical values from the standard normal distribution. An approximate 95% confidence interval for $\hat{\beta}_i$ is $\hat{\beta}_i - 1.96.\text{se}(\hat{\beta}_i)$ to $\hat{\beta}_i + 1.96.\text{se}(\hat{\beta}_i)$. Use can be made of differences between log-likelihood goodness-of-fit statistics as a means of determining the amount of evidence for selection. It is possible to compare observed and expected sample counts by using appropriately defined standardized residuals.

For the models being considered in the present chapter the log-likelihood goodness-of-fit statistics are minus twice the maximized log-likelihood and are therefore of the form

$$X_L{}^2 = 2\sum O_i \log_e(O_i/\hat{E}_i),$$

where O_i denotes the ith observed frequency, $\hat{E}_i$ denotes the ith expected frequency from an estimated model, and the summation is over all sample frequencies. If the expected frequencies are reasonably large (mostly five or more) then these statistics can be compared with critical values of the chi-squared distribution to assess the goodness-of-fit of models. Otherwise, it is only valid to assess differences between $X_L{}^2$ using the chi-squared distribution.

The standardized residual for the difference between an observed frequency O and the corresponding expected frequency $\hat{E}$ from a fitted model can be defined as $R = 2(O - \hat{E})/\sqrt{\hat{E}}$. Providing that most of the expected frequencies are reasonably large (five or more) the values of R should behave as approximately independent standard normal variables if the fitted model is correct.

7.7.1 Example 7.1 Nest selection by fernbirds

Consider again Harris' (1986) data on 24 nest sites selected by fernbirds (*Bowdleria puncta*) and 25 random sites from the same region of Otago, New Zealand (Table 2.4). A star plot comparison between the two samples, produced using the SOLO statistical package (BMDP, 1988), is shown in figure 7.1. Here, the star for each of the sites has three rays, with one for each of the variables (canopy height, distance to edge, and perimeter of clump), and the length of a ray is proportional to the relative value of the variable being used. The generally higher values of variables for the nest sites is shown quite clearly from the figure.

The means and standard deviations of the variables are shown in Table 2.4. The results of t-tests (with 47 degrees of freedom) to compare the difference between the means for nest sites and available sites are as follows: canopy height, $t = 3.86$, $p < 0.001$; distance to edge, $t = 3.23$, $0.01 > p > 0.001$; perimeter of clump, $t = 3.41$, $0.01 > p > 0.001$. The results of F-tests (with 23 and 24 degrees of freedom) to compare the sample variances, with the F-ratio

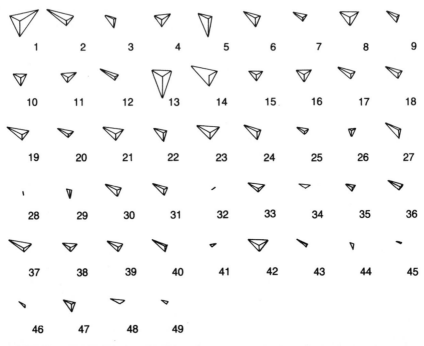

Figure 7.1   Star plots for the comparison of nest sites used by fernbirds and available sites. The vertical ray on each star represents the relative magnitude of the canopy height. Moving clockwise, the next ray represents the relative magnitude of the distance to the edge of the clump of vegetation, and the third ray represents the perimeter of the clump. The first 24 plots are for the nest sites and the last 25 plots are for the random sites.

being the nest site variance divided by the available site variance are as follows: canopy height, $F = 3.58$, $0.01 > p > 0.001$; distance to edge, $F = 0.61$, $p > 0.05$; and perimeter of clump, $F = 2.63$, $0.05 > p > 0.01$. These univariate tests show very clear evidence of a mean difference between nest and available sites, and evidence of a difference in the level of variation for two of the three variables. It is, of course, well known that the F-test as used here is rather sensitive to non-normality in the data and alternative more robust tests are available (Manly, 1986, p. 32). However, the F-tests serve quite well as simple indicators of variance differences.

   Multivariate tests can also be carried out. Box's likelihood ratio test to compare the two sample covariance matrices (Morrison, 1976, p. 252) shows a difference that is significant at the 1% level, while Wilk's lambda test to compare the two vectors of sample means (Morrison, 1976, p. 222) gives a result that is significant at the 0.1% level. The latter test involves the assumption that the two samples are from populations with the same covariance matrix, which must be doubted because of the result from the first test. However, there is clearly some evidence of significant differences between the nest sites and available sites in terms of both mean levels and the amount of variation.

Since the nest sites used were only a small, unknown fraction of the potential sites available, and sampling fractions are unknown, this example matches the small fraction used model for the estimation of a resource selection function, as discussed in section 7.8. Thus, equation (7.6) can be used to give the expected frequency of type i sites in the nest sample, with canopy height x_{i1}, distance to edge x_{i2}, and perimeter of clump x_{i3}, as

$$E\{u_i(2)\} = P_u(2).A_i.\exp(\beta_0 + \beta_1 x_{i1} + \beta_2 x_{i2} + \beta_p x_{i3})2. \qquad (7.12)$$

Time is taken as $t = 2$ years since the nests were used over this period of time. There are 49 of these expected frequencies, one for each of the 49 different sampled sites, all of which have different sets of X values. The corresponding observed values are 1 for the units that are in the nest sample and 0 for the units in the available sample.

For the sample of available sites (unused sites at time $t = 0$ years), the expected frequency of type i sites is given by equation (7.7) to be

$$E\{\bar{u}_i(0)\} = P_{\bar{u}}(0).A_i. \qquad (7.13)$$

There are again 49 of these expected frequencies, one for each of the 49 different types of sites seen in the two samples. The corresponding observed frequencies are 1 for the sites that were in the available sample and 0 for the nest sites.

As discussed in section 7.4, to fit the model to the data it is necessary to set $P_u(2) = 1$, which means that the parameter $P_{\bar{u}}(0)$ becomes the sampling probability for the available sites divided by the sampling probability for the used sites. It is also necessary to set $\beta_0 = 0$ since it cannot be estimated. The parameters that can be estimated are then A_1 to A_{49}, $P_{\bar{u}}(0)$, and β_1 to β_3, of which only the last three are the important ones for assessing resource selection.

Using the program RSF, the estimated β values, with standard errors in parentheses, are found to be $\hat{\beta}_1 = 7.80$ (3.26), $\hat{\beta}_2 = 0.21$ (0.12), and $\hat{\beta}_3 = 0.88$ (0.48). The ratios of the estimates to their standard errors are therefore $7.80/3.26 = 2.39$, $0.21/0.12 = 1.73$, and $0.88/0.48 = 1.84$, respectively. Comparison with the standard normal distribution then shows that only $\hat{\beta}_1$ is significantly different from zero (at the 5% level).

The 'no selection' model for this set of data says that the estimated probability of a site being in the nest sample is $24/49 = 0.49$, and the probability of being in the available sample is $25/49 = 0.51$, given that the site appears in one of the samples. These estimated probabilities are simply the respective sample sizes divided by the total number of units seen. Thus for each of the 49 sites seen in both samples, the expected frequency for the nest sample is 0.49 and the expected frequency for the available sample is 0.51. There are then 98 expected frequencies altogether, with which the corresponding 98 observed frequencies can be compared.

The log-likelihood chi-squared value for the 'no selection' model is found to be 67.91, with 48 degrees of freedom. This cannot be used as a valid goodness-of-fit statistic because of the small expected frequencies. However, it can be compared with the log-likelihood chi-squared value of 40.48, with 45 degrees of freedom that is obtained for the model that is expressed by equations (7.12) and (7.13). The difference of 27.43 is very highly significant when compared with the chi-squared distribution with three degrees of freedom. There is therefore very strong evidence that the selection of nest sites was related to the three measured variables.

From equation (7.12) the estimated resource selection function is

$$w(x,t) = \exp\{7.80(\text{CANOPY}) + 0.21(\text{EDGE}) + 0.88(\text{PERIM})\}t,$$

with obvious abbreviations for the variables canopy height, distance to the edge of the clump of vegetation, and the perimeter of the clump. Table 7.1 shows the values that are obtained by this function after it has been scaled so as to make the average value 1 in the sample of available nest sites. This scaling, which is achieved by evaluating $w(x,t)$ for each of the 49 sampled sites and then dividing by the average of those in the available sample, has been made in order to ensure that the values are within a convenient range. Since only relative values of the resource selection function are important any scaling of this type is immaterial as far as interpreting results is concerned.

The scaled values from the selection function indicate that some of the sites that were used for nests had very much higher probabilities of use than the sites in the available sample, leading to a mean for the scaled selection probabilities of 1109.0 for the nest sites compared to 1 for the available sites. There is also much more variation for the nest sites, which have a standard deviation of 3718.3 compared with 1.9 for the available sites.

There are further calculations that could be done with this example. In particular, the effect of removing one or both of the variables EDGE and PERIM from the resource selection function could be examined. Square and product terms could also be added to the function. However, these possibilities will not be considered here.

7.7.2 Example 7.2 Selection of *Daphnia* by yellow perch

Table 7.2 shows plankton and yellow perch (*Perca flavescens*) stomach samples of *Daphnia publicaria* taken by Wong and Ward (1972) on five different days in 1969 from West Blue Lake in Manitoba. The investigators recorded the lengths of the *D. publicaria* in both samples, and considered the question of whether the predators were selective, and whether the selection changed with time.

An analysis of these data might reasonably begin with chi-squared tests to compare the frequencies of different sizes of *Daphnia publicaria* in the two types of sample taken on each of the five days, and also possibly to compare the

Table 7.1 Estimated values of the resource selection function for the fernbird data, after scaling to have a mean of 1 for the sample of available sites; the sites are in the same order as in Table 2.4

	Scaled resource selection function	
	Nest sites	Available sites
	15030.4	0.3
	21.5	0.3
	1.2	4.3
	7.4	0.0
	294.7	0.3
	7.6	1.6
	0.3	1.2
	19.0	0.0
	0.9	0.7
	3.1	0.1
	1.8	0.2
	0.8	0.3
	11021.8	3.4
	152.0	0.6
	1.1	1.1
	7.8	0.7
	1.5	0.0
	1.4	8.5
	6.8	0.1
	0.4	0.1
	9.8	0.0
	3.0	0.0
	17.6	1.2
	4.4	0.1
		0.0
Mean	1109.0	1.0
SD	3718.3	1.9
Minimum	0.3	0.0
Maximum	15030.4	8.5

frequencies in samples taken at different times. However, these tests are largely superfluous because of the clear differences between the plankton and stomach samples, at least for the first two days, and the fact that differences can be expected between the samples taken on different days because the *D. publicaria* and perch were growing in size. Therefore, this example will consider the estimation of resource selection functions only.

Table 7.2 Distributions of the lengths of *Daphnia publicaria* in plankton (P) and in the stomachs (S) of yellow perch fry in five samples taken on different days in 1969 from West Blue Lake, Manitoba[*]

Length (mm)	1 July		15 July		29 July		12 August		25 August	
	P	S	P	S	P	S	P	S	P	S
0.5–0.7	20	59	28	20	2	0	1	0	6	27
0.7–	22	84	49	40	11	12	2	0	2	42
0.9–	20	154	59	101	21	61	7	34	2	124
1.1–	18	138	62	126	33	95	9	127	0	138
1.3–	26	44	46	146	59	172	17	230	3	261
1.5–	24	10	33	60	31	233	28	241	12	303
1.7–	22	5	28	2	24	168	14	218	35	604
1.9–	24	0	33	5	22	78	12	218	63	606
2.1–	26	0	13	2	16	21	4	92	36	289
2.3–	16	0	13	2	11	9	6	34	15	193
2.5–	11	0	7	0	7	1	4	11	5	55
2.7–	7	0	7	0	2	0	1	5	0	58
2.9–	1	0	2	0	1	0	0	6	0	0
Total	237	494	380	504	240	850	105	1216	179	2710
Mean	1.56	1.03	1.39	1.23	1.58	1.56	1.66	1.70	1.94	1.82
SD	0.59	0.25	0.55	0.29	0.47	0.31	0.44	0.36	0.38	0.43

* Constructed from Figure 1 of Wong and Ward (1972).

This example is similar to the previous one in the sense that the universe of available resource units (*D. publicaria*) was very large, with only a small and unknown fraction of these being used (eaten). Also, there is no information available on sampling probabilities. Therefore, this is another situation where the small fraction used model applied. Of course, the available prey would have been changing continually over the study period of nearly two months. However, each of the five sample days gives a 'snapshot' of the resource selection at one point in time if the plankton sample on each day can be regarded as a random sample from the universe of *D. publicaria* that were available to the perch fry at that time. The approach that is used here therefore involves fitting a separate resource selection function to the data for each of the sampling days.

From preliminary modelling of the data, which will not be considered in detail here, it was decided to model the resource selection function with a cubic relationship and assume that for each of the sampling days the resource selection probability function was of the form

$$w^*(x,t) = \exp(\beta_0 + \beta_1 x + \beta_2 x^2 + \beta_3 x^3)t,$$

taking the selection time t = 1. For consistency, the same function was fitted to the data on all five days, although linear or quadratic functions give almost as good a fit for some of these days.

With this assumption, equation (7.6) for the expected number of *D. publicaria* in size class i in a stomach sample becomes

$$E\{u_i(1)\} = P_u(1).A_i.\exp(\beta_0 + \beta_1 x_i + \beta_2 x_i^2 + \beta_3 x_i^3), \tag{7.14}$$

while equation (7.7) for the expected frequency of the same size class in the plankton sample becomes

$$E\{\bar{u}_i(1)\} = P_{\bar{u}}(1).A_i. \tag{7.15}$$

For the first four sampling days there are 13 size classes, and 12 size classes on 25 August. These size classes consist of *D. publicaria* with lengths within 0.2 millimetre ranges, and the x_i values that are used in equations (7.14) and (7.15) can be taken to be the centres of these ranges, which are 0.6, 0.8,..., 3.0.

As discussed in section 7.4, β_0 is confounded with the sampling probability $P_u(1)$ in equation (7.14), and it is not possible to estimate both of these parameters. However, this problem is overcome by setting $\beta_0 = 0$. The sampling probabilities are also confounded with the A_i parameters, which can be overcome by setting $P_{\bar{u}}(1)=1$. The unknown parameters for the data from the first four sampling days are then A_1 to A_{13}, $P_u(1)$, and β_1 to β_3. For the last sampling day there are only 12 of the A_i parameters.

The models were estimated using the program RSF for each of the five sampling days, with the following results:

Samples taken on 1 July

Log-likelihood chi-squared value for the 'no selection' model = 328.26 with 12 degrees of freedom (df); log-likelihood chi-squared for the cubic model = 10.79 with 9 df. The difference of 317.47 with 3 df is very highly significant. The chi-squared value for the cubic model is not significantly large at the 5% level, which indicates that the model is a good fit. The estimated resource selection function (with standard errors in parenthesis below the corresponding estimates) is

$$w(x,1) = \exp(15.490x - 8.936x^2 + 0.705x^3).$$
$$\quad\ (8.763)\quad\ (7.852)\quad\ (2.229)$$

In this case it is apparent that a quadratic or linear function in x would provide a model that fits well since the coefficients of x^2 and x^3 are not at all significant.

Samples taken on 15 July

Log-likelihood chi-squared value for the 'no selection' model = 171.06 with 12 df; log-likelihood chi-squared for the cubic model = 39.01 with 9 df. The difference of 132.05 with 3 df is very highly significant. The chi-squared value for the cubic model is very significantly large, which indicates that the model is a rather poor fit. The estimated resource selection function is

$$w(x,1) = \exp(10.120x - 4.253x^2 + 0.964x^3).$$
$$(5.121) \quad (4.011) \quad (0.997)$$

Again, it is apparent that a quadratic or linear function in x would provide a model that fits well since the coefficient of x^3 is not at all significant.

Samples taken on 29 July

Log-likelihood chi-squared value for the 'no selection' model = 90.16 with 12 df; log-likelihood chi-squared for the cubic model = 19.29 with 9 df. The difference of 70.87 with 3 df is very highly significant. The chi-squared value for the cubic model is significantly large at the 5% level, which indicates that the model is a poor fit. The estimated resource selection function is

$$w(x,1) = \exp(-3.957x + 5.782x^2 - 1.807x^3).$$
$$(6.337) \quad (4.126) \quad (0.862)$$

Here the coefficient of x^3 is more than two standard errors from zero, which indicates that this term is needed in the model.

Samples taken on 12 August

Log-likelihood chi-squared value for the 'no selection' model = 36.10 with 12 df; log-likelihood chi-squared for the cubic model = 22.69 with 9 df. The difference of 13.41 with 3 df is significantly large at the 1% level. The chi-squared value for the cubic model is also significantly large at the 1% level, which indicates that the model is a poor fit. The estimated resource selection function is

$$w(x,1) = \exp(16.010x - 7.617x^2 + 1.135x^3).$$
$$(6.578) \quad (3.704) \quad (0.670)$$

This is another case where the coefficient of x^3 is not significant.

Samples taken on 25 August

Log-likelihood chi-squared value for the 'no selection' model = 80.79 with 11 df; log-likelihood chi-squared for the cubic model = 9.44 with 8 df. The difference of 71.35 with 3 df is very significantly large. The chi-squared value for the cubic model is not significantly large, which indicates that the cubic model is a good fit. The estimated resource selection function is

$$w(x,1) = \exp(31.72x - 20.85x^2 + 4.146x^3).$$
$$(4.72) \quad (3.11) \quad (0.635)$$

Here the coefficient of x^3 is more than six standard errors from zero, which makes it very highly significant.

Chi-squared goodness-of-fit tests are valid with this example because most of the observed and expected frequencies are fairly large. The lack of fit of the model in three out of five cases is therefore rather unsatisfactory, but it may be caused by the pseudoreplication that is implied by taking stomach samples of *D. publicaria*, rather than the inadequacy of the model. The point here is that the samples of used resource units were presumably cluster samples (of *D. publicaria* in a few stomachs) rather than random samples of individual eaten *D. publicaria*. The use of cluster sampling rather than random sampling can easily account for the correct model giving an apparently poor fit to data.

As discussed briefly in section 2.4, one way to make some allowance for a poor fitting model, where this is thought to be due to cluster sampling rather than to the model being wrong, involves assuming that the variances of all the observed sample frequencies are inflated by the same multiplicative heterogeneity factor, H, which can be estimated by $\hat{H}$, the chi-squared goodness-of-fit statistic for the fitted model, divided by its degrees of freedom. The standard errors obtained on the assumption that observed sample frequencies have Poisson distributions can then be adjusted to allow for the heterogeneity factor by multiplying them by $\sqrt{\hat{H}}$ (Manly, 1990, p. 39).

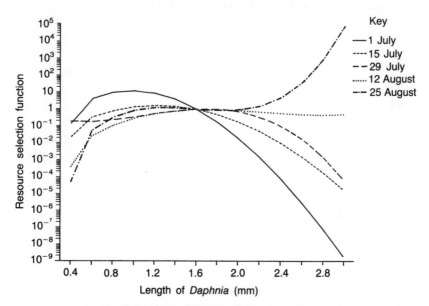

Figure 7.2   Estimated resource selection functions for *Daphnia publicaria* predated by perch fry in West Blue Lake, over the period from 1 July to 25 August. The resource selection functions estimated for each of five days have been scaled to have the value 1 for *D. publicaria* with a length of 1.6 mm. A logarithmic scale is used on the vertical axis to accommodate the large range of values.

For example, consider the data for 29 July. Here the log-likelihood chi-squared model for the fitted data is 19.29 with 9 degrees of freedom. Hence $\hat{H}=$ 19.29/9 = 2.14, which means that the nominal standard errors for the β estimates should be multiplied by $\sqrt{2.14}$ = 1.46 to allow for cluster sampling.

It must be stressed that this type of adjustment is not valid unless the assumptions that are being made are correct. In particular, it is not applicable if the model is a poor fit simply because it is the wrong type of model. For this reason, a better way of handling cluster sampling in the context of the present example would involve estimating a separate resource selection function for each fish stomach available and using the variation between stomachs to estimate the variance of sampling errors.

Although the cubic model does not always fit the data it does produce interesting approximations for the resource selection functions. These approximations are plotted for each of the sampling days in figure 7.2 after they have been scaled so that in each case the values are exactly 1 for *D. publicaria* in the 1.6 mm size class. The enormous range of values for these scaled resource selection functions has made it necessary to plot their logarithms rather than the values themselves. It is interesting to see how the functions change in a systematic way from 1 July to 25 August. On 1 July it seems that smaller prey were preferred, and large prey hardly eaten at all. However, by 25 August the larger prey were being used far more than the smaller prey.

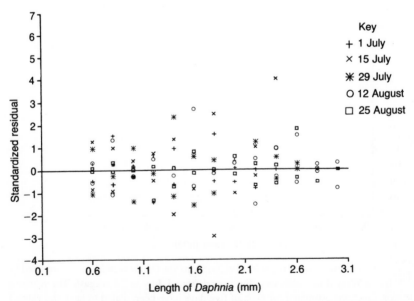

Figure 7.3 Standardized residuals, $R = (O-\hat{E})/\sqrt{\hat{E}}$, for the models fitted to the data on the selection of *Daphnia publicaria* by perch fry.

Figure 7.3 shows standardized residuals of the form $R = (O-\hat{E})/\sqrt{\hat{E}}$, where O represents the observed number of *D. publicaria* in a size class in one of the samples, with a corresponding expected frequency of $\hat{E}$ from the fitted model. It will be recalled that these standardized residuals should behave like approximately standard normal variates. In fact they do generally have this appearance, except for one very extreme positive value corresponding to a higher than expected frequency for *D. publicaria* with a length in the range 2.3–2.5 mm in the stomach sample on 15 July, and one fairly extreme negative value corresponding to a much lower than expected frequency for *D. publicaria* with a length in the range 1.7–1.9 mm in the sample on 15 July. Since the first of these standardized residuals is based on a very small expected frequency it does not have to be taken too seriously so that, on the whole, the residuals from the fitted model are reasonable.

As with the example 7.1, there is more that could be done with the data, including the investigation of alternative equations for the resource selection function. However, these possibilities will not be followed up here.

7.7.3 Example 7.3 Selection of corixids by minnows

For a third example, consider Popham's (1944) study of the use of corixids as food by minnows (*Phoxinus phoxinus*) for which the data are shown in Table 2.5. This was the subject of example 2.5, and it may be recalled that Popham sampled the corixid in a pond before introducing minnows, and then daily from three to nine days of predation. No information is available about sampling fractions for this example since Popham endeavoured to obtain about 150 corixids in each of his daily samples, irrespective of the number present in the population.

There are three obvious models that can be entertained to account for the sample data. The simplest of these is the 'no selection' model, which says that the population proportions of the nine types of corixid were the same at all the sample times, and the sample differences in proportions were just due to random sampling effects. For this model the expected frequencies of the nine types of corixid for each sample are just the sample total allocated out according to the proportions found with all samples lumped together, as in a standard chi-squared test for no association between row and column categories in a two-way contingency table. On this basis, the log-likelihood goodness-of-fit statistic is $X_L^2 = 174.59$, with 56 degrees of freedom. This is very significantly large, indicating that the model is a very poor fit to the data.

The second model that can be entertained says that the resource selection probability function is dependent on a species effect and a colour effect, with no interaction between these. This can be achieved by taking $X_1 = 1$ for *Sigara venusta*, or otherwise 0, $X_2 = 1$ for *S. praeusta*, or otherwise 0, $X_3 = 1$ for light corixids, or otherwise 0, and $X_4 = 1$ for medium corixids, otherwise 0. The 'standard' corixid is then a dark *S. distincta*, for which all X values are zero.

Since only samples of uneaten corixids are available and sampling probabilities are unknown, the unused units only model is appropriate. Hence the form of the resource selection probability function can be assumed to be given by equation (7.9), which with four X variables becomes

$$w^*(x_i, t) = 1 - \exp\{(\beta_0 + \beta_1 x_{i1} + \beta_2 x_{i2} + \beta_3 x_{i3} + \beta_4 x_{i4})t\}.$$

The expected number of type i corixids in the sample taken before the minnows were introduced (at time 0) is then

$$E(a_i) = P_a.A_i,$$

and the expected number in the sample taken t days later is

$$E\{\bar{u}_i(t)\} = P_{\bar{u}}(t).A_i.\exp\{(\beta_0 + \beta_1 x_{i1} + \beta_2 x_{i2} + \beta_3 x_{i3} + \beta_4 x_{i4})t\},$$

for i = 1 to 9 and t = 3 to 9.

As discussed in section 7.5, the last equation shows that any change in β_0 can be compensated for by changing the $P_{\bar{u}}(t)$ values. Hence to identify the other parameters β_0 must be set equal to zero. It is also necessary to set one of the sampling probabilities equal to one in order to identify the A_i parameters. The other sampling probabilities that are estimated then become sampling probabilities relative to the one that has been fixed. What this means in the end is that the parameters that can be estimated for this model are as follows:

(a) There are sampling probabilities for seven of the eight samples, relative to the sampling probability for the other sample. Choosing to set $P_a = 1$, this means that the parameters $P_{\bar{u}}(3)$ to $P_{\bar{u}}(9)$ can be estimated.
(b) There are nine parameters A_1 to A_9 that account for the initial number of corixids available in each of the species-colour classes, where these can be estimated by equating the total observed and expected numbers of corixids in each of the classes for all samples.
(c) There are four parameters β_1 to β_4 in the resource selection probability function. These are of course the important parameters for inferences.

Estimated β parameters obtained using the program RSF are shown in Table 7.3. The ratios $\hat{\beta}_i / se(\hat{\beta}_i)$ in the last column of the table indicate that the estimates of β parameters other than β_2 are significantly different from zero. Also, the log-likelihood goodness-of-fit statistic of $X_L^2 = 59.06$ with 52 degrees of freedom (the number of observations, 72, minus the number of estimated parameters, 20) is not significantly large at the 5% level in comparison with the chi-squared distribution. The model is therefore a good fit. Note that this chi-squared test is valid for this example because most of the observed and expected frequencies are reasonably large.

The third model that can be entertained for the data is one which says that the level of selection varies with all nine different types of corixid. This can be fitted by defining eight dummy variables to be $X_1 = 1$ for medium *S. venusta*, or otherwise 0, $X_2 = 1$ for dark *S. venusta*, or otherwise 0, and so on up to $X_8 = 1$

Table 7.3 Estimates of parameters of a resource selection probability function for the experiment on predation of corixids by minnows: model with additive species and colour effects

Parameter	β Estimate	Standard error	Estimate/std error
Sigara venusta	0.105	0.038	2.76
Sigara praeusta	0.060	0.050	1.20
Light corixids	−0.156	0.035	−4.45
Medium corixids	0.093	0.017	5.47

for dark *S. distincta*, or otherwise 0. The 'standard' corixid is then light *S. venusta*, for which all X values are zero. Any other of the species-colour combinations would have served just as well as the standard.

For the same reasons as applied with the second model β_0 cannot be estimated but can be set equal to zero, and P_a can be set equal to 1 so that the other sampling probabilities can be estimated relative to this 'standard' sampling probability. There is, however, a new complication that emerges with the third model because dark *S. distincta* were not seen at all after the introduction of minnows. This means that any model will fit the data on these corixids exactly providing that it makes the probability of one of them surviving from 19 to 22 September equal to zero. Therefore, the maximum likelihood estimator of β_8 must be negative infinity. This fact upsets the iterative process for finding maximum likelihood estimates since converging to infinity is not possible. A way round this difficulty involves setting $\beta_8 = -\infty$, and omitting all the data on dark *S. distincta* when estimating the other parameters of the model. The unknown parameters are then the relative sampling probabilities $P_{\bar{u}}(3)$ to $P_{\bar{u}}(9)$, the parameters A_1 to A_8 that account for the initial number of corixids available in each of the eight species–colour classes other than *S. distincta*, and β_1 to β_7.

Table 7.4 provides a summary of the estimates obtained for the coefficients of the variables X_1 to X_7 when the third model is fitted using the RSF program. The log-likelihood goodness-of-fit statistic is 53.69, with 48 degrees of freedom, which indicates a good fit.

The three different models can now be compared in an analysis of deviance table, which shows log-likelihood chi-squared goodness-of-fit values and their differences. This is provided in Table 7.5, from which it can be seen that:

(a) the no selection model provides a very poor fit to the data since the X_L^2 value is so large;
(b) there is a very highly significant reduction in the X_L^2 value when additive effects of species and colour are allowed for; and
(c) permitting each of the species–colour combinations to have a different level of selection leads to an insignificant reduction in the X_L^2 value.

In other words, the second model that was considered for the data seems best.

Table 7.4 Estimates of parameters of a resource selection probability function for the experiment on predation of corixids by minnows: model with different selection for each species–colour combination

Species	Colour	β Estimate	Standard error	Estimate/std error
S. venusta	Light	0	0	–
S. venusta	Medium	0.244	0.035	6.95
S. venusta	Dark	0.156	0.038	4.15
S. praeusta	Light	–0.047	0.091	–0.52
S. praeusta	Medium	0.205	0.051	4.01
S. praeusta	Dark	0.084	0.124	0.68
S. distincta	Light	–0.271	0.204	–1.33
S. distincta	Medium	0.163	0.052	3.14
S. distincta	Dark	$-\infty$	–	–

Table 7.5 Analysis of deviance table for three models fitted to the data from the experiment on predation of corixids by minnows

Model	X_L^2	df	X_L^2	df
No selection	174.59*	56		
			115.53*	4
Selection on species and colour, with independent effects	59.06	52		
			5.36	4
Different selection for each species–colour combination	53.70	48		

*Significantly large at 0.1% level

Having decided that the second model gives a reasonable fit to the data overall, it is possible to calculate standardized residuals of the form $R = (O - \hat{E})/\hat{E}$, where O and $\hat{E}$ are observed and expected frequencies of the number of corixids in one of the species–colour categories, and see whether these appear to be approximately independent standard normal variables. This can be assessed from figure 7.4, which shows these standardized residuals plotted against the sample numbers. Here all of the standardized residuals appear reasonable except for the large value of about five which occurs because the observed frequency of 58 dark S. venusta has an expected frequency of only 29.7. Clearly the observed frequency is anomalous. However, it is a genuine value so that a fair conclusion is that the fitted model is generally reasonable but there may have been some problem with the sampling for the third sample.

Finally, having chosen an appropriate model for the data it is possible to determine what the estimated resource selection function has to say about the

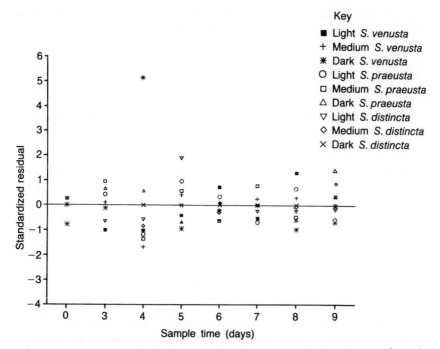

Figure 7.4   Standardized residuals for a log–linear model fitted to the data from the experiment on predation of corixids by minnows.

selection of corixids by minnows. The estimated resource selection probability function is

$$\hat{w}^*(x,t) = 1 - \exp\{(\beta_0 + 0.105x_1 + 0.060x_2 - 0.156x_3 + 0.093x_4)t\},$$

where β_0 cannot be estimated. This means that the estimated probability of a resource unit not being used in one day is proportional to

$$\hat{\phi}(x,1) = \exp\{(0.105x_1 + 0.060x_2 - 0.156x_3 + 0.093x_4)\}.$$

Values from this function are shown in Table 7.6. It appears that *S. venusta* survived better than *S. praeusta*, which in turn survived better than *S. distincta*, while medium coloured corixids survived better than dark corixids, which

Table 7.6   Estimated relative survival rates per day for different types of corixid

Species	Light	Medium	Dark
S. venusta	0.95	1.22	1.11
S. praeusta	0.91	1.15	1.06
S. distincta	0.86	1.10	1.00

survived better than light corixids. In other words, the corixids that were favoured by the minnows tended to be medium coloured and of the species *S. distincta*.

7.8 RECOMMENDATIONS

Three models (general, small fraction used, and unused units only) have been discussed in this chapter, and it is convenient to end with a summary of recommendations for their use:

(a) *The general model.* If sampling probabilities or relative sampling probabilities are known, then the resource selection probability function of equation (7.4) can be estimated from samples of available, used and/or unused resource units, with equations (7.1) to (7.3) being used to calculate expected frequencies for units with different values for the variables X that are measured on them. The computer program RSF can be used for the calculations, or alternatively, a general program for maximum likelihood estimation. This approach to estimation can also be used if samples of both used and unused units are available, with an appreciable proportion of the available units being used by the end of the study period.

(b) *The small fraction used model.* If the proportion of resource units used during the study period is small, and relative sampling probabilities are unknown, then the resource selection probability function can be approximated by equation (7.8) with $\hat{\beta}_0$ set equal to 0, and the expected sample frequencies of different types of used and unused units by equations (7.6) and (7.7). The resource selection probability function can then only be estimated to within an arbitrary multiplicative constant, but this still allows the relative attractiveness of different units to be determined. The calculations can be carried out using a computer program for log-linear modelling or RSF.

(c) *The unused units only model.* If only samples of unused resource units are available and relative sampling probabilities are unknown then the resource selection probability function can be approximated using equation (7.9), and the expected sample frequencies of different types of unit by equation (7.10). It is then only possible to estimate some arbitrary multiple of the probability of a resource unit not being used by time t, but this still allows the sampled resource units to be ranked in order of desirability. The calculations can be carried out using a computer program for log-linear modelling, or RSF.

Exercises

(1) In a study carried out in Marley Wood, near Oxford, England, Sheppard (1951) collected two samples of live snails (*Cepaea nemoralis*) and seven samples of broken shells close to a thrush 'anvil'. The results obtained are shown in Table 7.7, with the snails classified according to the colour of the

shells (pink and brown or yellow), and the nature of the sample (broken shells or live snails).

Table 7.7 Colour composition of samples of live (L) and broken (B) *Cepaea nemoralis* from Marley Wood, Oxford, with days counted from 6 April, 1950

	Day								
	5	8	17	24	31	43	46	50	50
Type of sample	B	L	B	B	B	B	B	B	L
Pink and brown	4	250	10	21	25	16	6	12	147
Yellow	3	80	7	11	9	3	1	2	57

On Day 1 (6 April, 1950) the thrush 'anvil' was cleared of broken shells, so that the sample of broken shells collected on day 5 was of snails taken by thrushes over four days, the sample of broken shells collected on day 17 was a sample of snails taken by thrushes over 12 days, and so on. The two samples of live snails were of the available snails at the sample times.

(a) Carry out a chi-squared goodness-of-fit test, using the log-likelihood test statistic, for the 'no selection' model that makes the expected proportion of pink and brown snails the same in each of the samples. (This is a regular chi-squared test for the independence of the two categories of the colour of the shell and the sample day.)

(b) fit a resource selection model that allows for a constant preference for one colour of shell, assuming that the universe of available snails is so large that it is effectively unchanged by the thrush predation, so that the small fraction used model can be used. That is to say, assume that the expected frequency of uneaten type i snails in a sample is $E\{\bar{u}_i(t)\} = P_{\bar{u}}(t).A_i$, and the expected frequency of broken shells is given by $E\{u_i(t)\} = P_u(t).A_i.\exp(\beta_0 + \beta_1 x_i)t$, where $x_i = 0$ for pink and brown snails, and 1 for yellow snails, and β_0 has to be set to zero because it cannot be estimated. Because of the way that thrush anvils were cleared when collecting broken shells, the times t used with the samples of broken shells is best taken to be the time since the previous collection rather than the time since the start of the study.

(c) Sheppard (1951) was interested in the possibility that the preference of the thrushes was changing during the study period. Examine this by estimating the model of part (b) for the first four and the last five samples separately. (See also the log-linear model analysis given by example 4.2 of Manly (1985).)

(2) Recall that example 7.3 was concerned with a study carried out by Popham (1944) on the selection of corixids by minnows. Another study by Popham (1966), on the selective predation of corixids (*Vermicorixa nigrolineata*) by

three spined sticklebacks (*Gasterosteus aculeatus*), is the subject of the present exercise. On 22 and 23 August, 1950, Popham introduced 1100 specimens of *V. nigrolineata* into a pond where this species was not previously present. During the transfer, 84 corixids were held back, to provide a random sample of 7.09% of the population. Popham observed the introduced corixids being attacked by sticklebacks, and sampled the corixids again on 26 August and 29 August to assess the changes to the population that predation may have caused, in terms of the proportions of pale, medium and dark corixids. He also estimated the size of the corixid population on 26 August and 29 August by mark recapture methods. The results of the experiment are summarized in Table 7.8.

Table 7.8 Samples of *Vermicorixa nigrolineata* taken from an artificial population set up on 22 and 23 August, 1950

Sample date	Time (days)	Colour of corixids			Estimated population size
		Pale	Medium	Dark	
August 22/23	0	31	27	26	1184
August 26	3.5	0	22	29	400
August 29	6.5	3	39	72	231

Analyse these results using the general model. Treat the sample taken from the corixids introduced on 22 and 23 August as being a random sample from the universe of available resource units, with a sampling probability of $(31+27+26)/1184 = 0.071$; treat the sample taken on 26 August as a random sample of unused resource units at time 3.5, with the sampling probability $(0+22+29)/400 = 0.128$; and treat the sample taken on 29 August as a random sample of unused resource units at time 6.5 days, with sampling probability $(3+39+72)/231 = 0.494$. Use two appropriate dummy variables to account for a different preference of the sticklebacks for the different colours of prey.

8 Estimating a resource selection function from two samples of resource units using logistic regression and discriminant function methods

When a study of resource selection involves the collection of only two samples it is possible to simplify the estimation of a resource selection probability function. In particular, under certain conditions the estimation can be carried out using logistic regression or discriminant function analysis. The purpose of the present chapter is therefore to discuss the application of these two methods, and relate them to the models that have been described in the previous chapter.

8.1 LOGISTIC REGRESSION

A basic result that can often be used to justify logistic regression with samples of resource units comes from the following relationship between the Poisson and binomial distributions. Suppose that the observations in a sample are classified into I different classes, where Y_{i1}, the count of the number of observations in class i, is a random value from a Poisson distribution with mean μ_{i1}. Suppose also that at the same time, the observations in a second sample are also classified into the same I classes, where Y_{i2}, the count of the number of observations in class i, is an independent Poisson random variable with mean μ_{i2}. Then the distribution of Y_{i1}, conditional on the sum $Y_{i1} + Y_{i2}$ being equal to a particular value n_i, is a binomial distribution with mean

$$E(Y_{i1}|Y_{i1}+Y_{i2}=n_i) = n_i\tau_i, \qquad (8.1)$$

and variance

$$Var(Y_{i1}|Y_{i1}+Y_{i2}=n_i) = n_i\tau_i(1-\tau_i), \qquad (8.2)$$

where $\tau_i = \mu_{i1}/(\mu_{i1}+\mu_{i2})$ can be interpreted as the probability of a type i observation being in the first sample rather than in the second sample, given that it is in one of the two samples (McCullagh and Nelder, 1989, p. 101).

The importance of this result in the present context comes about because it is possible to identify the two samples just considered as samples of resource units of different types. In that case, if τ_i can be thought of as a logistic function of X variables measured on the resource units then the parameters of a resource selection probability function can be estimated by logistic regression.

Three situations must be considered, corresponding to where the two samples to be analysed are (a) a sample of available and a sample of used units at some time t, (b) a sample of available and a sample of unused units at some time t, and (c) a sample of unused and a sample of used units at some time t. These cases will now be considered in turn.

8.1.1 Samples of available and used resource units

As before, the population of available units before any units have been used is considered to consist of i different groups, with the A_i units in group i having the values $x_i = (x_{i1},x_{i2}, \ldots ,x_{ip})$ for the variables X_1 to X_p. The situation that will now be considered occurs when this population is sampled in such a way that every available resource unit has the same probability P_a of being found in a random sample of available units, and every unit that has been used after a certain period of selection has the same probability P_u of being found in a sample of used units.

Suppose that there are a_i type i units in the available sample and u_i type k units in the used sample. Suppose also that the population of resource units is large and the sampling probabilities are small, so that a_i and u_i will approximately have independent Poisson distributions. Then the mean value of a_i will be

$$E(a_i) = P_aA_i, \qquad (8.3)$$

and the mean of u_i will be

$$E(u_i) = P_uA_iw^*(x_i), \qquad (8.4)$$

where $w^*(x_i)$ is the probability of use for a type i unit, which is, of course, the resource selection probability function.

It is convenient at this point to assume that the resource selection probability function takes the particular form

$$w^*(x) = \exp(\beta_0 + \beta_1x_1 + \ldots + \beta_px_p), \qquad (8.5)$$

where the argument of the exponential function should be negative. This is equation (7.5) for the model in the previous chapter that assumes that there are a small proportion of resource units used, with a selection time of t = 1. It follows using equations (8.1) and (8.2) that the distribution of u_i, conditional on the total number of type i units in both samples being equal to $n_i = a_i + u_i$, is a binomial distribution with mean $n_i \tau_i$ and variance $n_i \tau_i (1-\tau_i)$, where

$$\tau_i = \frac{\exp\{\log_e(P_u/P_a) + \beta_0 + \beta_1 x_{i1} + \ldots + \beta_{ip} x_{ip}\}}{1 + \exp\{\log_e(P_u/P_a) + \beta_0 + \beta_1 x_{i1} + \ldots + \beta_{ip} x_{ip}\}}. \tag{8.6}$$

This is a logistic regression equation in which the parameter β_0 is modified to $\log_e(P_u/P_a) + \beta_0$ to allow for available and used resource units being sampled with different probabilities. Hence the resource selection probability function can be estimated using any standard computer program for logistic regression when the data available come from a sample of used and a sample of available units. The dependent variable in the logistic regression is the number of used units of type i out of the total number of units of type i sampled, for i = 1 to I.

The fact that the constant in the logistic regression is $\log_e(P_u/P_a) + \beta_0$ means that if the ratio P_u/P_a is known then the parameter β_0 in the resource selection probability function can be estimated by subtracting $\log_e(P_u/P_a)$ from the estimated constant in the logistic regression equation. On the other hand, if P_u/P_a is not known then β_0 cannot be estimated. This agrees exactly with the situation for the small fraction used model as discussed in Chapter 7. If β_0 cannot be estimated then it is still possible to estimate the resource selection function

$$w(x) = \exp(\beta_1 x_1 + \ldots + \beta_p x_p) \tag{8.7}$$

and use this to compare resource units.

8.1.2 Samples of available and unused resource units

Suppose next that what is available for estimating a resource selection probability function is a sample of available resource units and an independent sample of unused resource units. The resource selection probability function can then conveniently be approximated by

$$w^*(x) = 1 - \exp(\beta_0 + \beta_1 x_1 + \ldots + \beta_p x_p), \tag{8.8}$$

where the argument of the exponential function should be negative. This is equation (7.9) for the unused units only model of the previous chapter with a selection time of t = 1.

The expected frequency of type i resource units in the sample of available units is given by equation (8.3), while the expected frequency for these units in the unused sample becomes

$$E(\bar{u}_i) = P_{\bar{u}} A_i \exp(\beta_0 + \beta_1 x_{i1} + \ldots + \beta_p x_{ip}), \tag{8.9}$$

where $P_{\bar{u}}$ is the sampling probability for unused resource units. Assuming that these sample counts have independent Poisson distributions, and using equations (8.1) and (8.2), it is then found that the distribution of the number of unused type i resource units, conditional on the total number of type i units in both samples being n_i, has a binomial distribution with mean $\tau_i n_i$ and variance $\tau_i(1-\tau_i)n_i$ where

$$\tau_i = \frac{\exp\{\log_e(P_{\bar{u}}/P_a) + \beta_0 + \beta_1 x_{i1} + \ldots + \beta_{ip}x_{ip}\}}{1 + \exp\{\log_e(P_{\bar{u}}/P_a) + \beta_0 + \beta_1 x_{i1} + \ldots + \beta_{ip}x_{ip}\}}. \tag{8.10}$$

This is a model that can be fitted using logistic regression taking the number of type i units in the unused sample as the dependent variable. Since the constant term in the logistic regression equates to $\log_e(P_{\bar{u}}/P_a) + \beta_0$, it follows that β_0 can be estimated only if the ratio of sampling probabilities is known.

8.1.3 Samples of unused and used resource units

The situation where there is a sample of unused units and a sample of used units can also be handled by logistic regression. In this case, the expected number of type i units in the sample of unused units will be of the form $E(\bar{u}) = P_{\bar{u}}A_i\{1-w^*(x_i)\}$, and the expected number of the same type in the sample of used units is $P_u A_i w^*(x_i)$, where $P_{\bar{u}}$ and P_u are sampling probabilities, A_i is the number of available type i units, and $w^*(x_i)$ is the resource selection probability function. It therefore follows, using equations (8.1) and (8.2), that the distribution of the number of used units of type i, conditional on the total number of units of type i in both samples being n_i, is binomial with mean $n_i\tau_i$ and variance $n_i\tau_i(1-\tau_i)$, where

$$\tau_i = \frac{P_u A_i w^*(x_i)}{P_{\bar{u}}A_i\{1 - w^*(x_i)\} + P_u A_i w^*(x_i)},$$

$$= \frac{(P_u/P_{\bar{u}})w^*(x_i)/\{1 - w^*(x_i)\}}{1 + (P_u/P_{\bar{u}})w^*(x_i)/\{1 - w^*(x_i)\}}.$$

The last equation defines a logistic regression by setting

$$(P_u/P_{\bar{u}})w^*(x_i)/\{1 - w^*(x_i)\} = \exp(\beta_0 + \beta_1 x_{i1} + \ldots + \beta_p x_{ip}),$$

in which case

$$\tau_i = \frac{\exp(\beta_0 + \beta_1 x_{i1} + \ldots + \beta_{ip}x_{ip})}{1 + \exp(\beta_0 + \beta_1 x_{i1} + \ldots + \beta_{ip}x_{ip})}, \tag{8.11}$$

and

$$w^*(x_i) = \frac{\exp\{\log_e(P_{\bar{u}}/P_u) + \beta_0 + \beta_1 x_{i1} + \ldots + \beta_p x_{ip}\}}{1 + \exp\{\log_e(P_{\bar{u}}/P_u) + \beta_0 + \beta_1 x_{i1} + \ldots + \beta_p x_{ip}\}}. \tag{8.12}$$

Thus, if a logistic regression is carried out, using the number of used type i units out of the total number of type i units as the dependent variable, then this will produce estimates of the β parameters in equation (8.11). An estimated resource selection probability function can then be obtained by substituting these estimates into equation (8.12), providing that the ratio of sampling probabilities $P_{\bar{u}}/P_u$ is known.

If the ratio of sampling probabilities is not known and cannot be estimated then it is not possible to estimate the resource selection probability function, or even this function multiplied by some unknown constant. The best that can be done is to arbitrarily set $P_{\bar{u}}/P_u = 1$ in equation (8.12) and recognize that the estimated function thereby obtained is an index of selectivity in the sense that if resource units are ranked in order using this function then they are being placed in the same order as they would be if the ratio of sampling probabilities were known.

8.1.4 Example 8.1 Nest selection by fernbirds

The study on nest selection by fernbirds (*Bowdleria puncta*) that produced the data shown in Table 2.4 has been discussed in example 2.4, and the data have been analysed in terms of a log-linear model in example 7.1. In fact, the estimated resource selection function that was obtained in example 7.1 can be found just as well by using logistic regression.

There is a sample of available resource units (random points in the study region) and a sample of used resource units (nest sites). Sampling fractions are unknown, but are clearly very small. A logistic regression was carried out using the SOLO statistical package (BMDP, 1988), with the dependent variable being 0 for available sites and 1 for nest sites, and the three variables canopy height, distance to edge, and perimeter of clump used as predictor variables. This produced the fitted equation

$$\hat{\tau} = \frac{\exp\{-10.73+7.79(\text{CANOPY})+0.21(\text{EDGE})+0.88(\text{PERIM})\}}{1 + \exp\{-10.73+7.79(\text{CANOPY})+0.21(\text{EDGE})+0.88(\text{PERIM})\}},$$

with obvious abbreviations for the variables. This corresponds to equation (8.6), so that the resource selection probability function would be estimated to be

$$\hat{w}^*(x)=\exp\{-10.73-\log_e(P_u/P_a)+7.79(\text{CANOPY})+0.21(\text{EDGE})+0.88(\text{PERIM})\}$$

if the ratio of sampling probabilities P_u/P_a were known. Since this ratio is not known, all that can be estimated is the resource selection function that is obtained by omitting the terms $(-10.73 - \log_e(P_u/P_a))$ from the last equation.

The estimated resource selection function is then exactly the same as the one obtained in example 7.1. The estimated standard errors of the coefficients of the predictor variables that are output by SOLO are also the same as those found from the log-linear model, as are the chi-squared goodness-of-fit statistics. In short, the log-linear model and the logistic regression model for these data are to all intents and purposes equivalent.

8.2 LINEAR DISCRIMINANT FUNCTION ANALYSIS

Linear discriminant function analysis can be thought of as a way to estimate a resource selection probability function for cases where the variables that are measured on resource units have multivariate normal distributions for the universes that are sampled. To see this, it is useful to begin by reviewing discriminant function analysis in terms of selection on a population.

Suppose that a population has a p variable multivariate normal distribution with a mean vector μ_1 and covariance matrix Σ, and that the members of the population are selected to form a second population in such a way that the probability of an individual with the values $x' = (x_1, x_2, \ldots, x_p)$ being selected takes the form

$$\Omega(x) = \exp(\beta_0 + \beta_1 x_1 + \ldots + \beta_p x_p), \tag{8.13}$$

where $\beta' = (\beta_1, \beta_2, \ldots, \beta_p)$. Then the second population will also have a multivariate normal distribution, with the vector of means being $\mu_2 = \mu_1 + \Sigma\beta$ and the covariance matrix still being Σ (Manly, 1985, p. 68). It follows that the β values other than β_0 in equation (8.13) are given by the equation

$$\beta = \Sigma^{-1}(\mu_2 - \mu_1), \tag{8.14}$$

which means that they are also the coefficients in Fisher's linear discriminant function, which is

$$L(x) = \beta_1 x_1 + \ldots + \beta_p x_p \tag{8.15}$$

(Seber, 1984, p. 109).

The value of β_0 can be related to the proportion of the first population that is selected to become part of the second population. This is the expected (mean) value of $\Omega(x)$ in the first population which is, from equation (8.13),

$$E\{\Omega(x)\} = \exp(\beta_0).E\{\exp(\beta' x)\}.$$

Here $E\{\exp(\beta' x)\}$ is the moment generating function of the first multivariate normal distribution with mean μ_1 and covariance matrix Σ, which is well known to be $\exp(\beta'\mu_1 + \frac{1}{2}\beta'\Sigma\beta)$. Hence

$$E\{\Omega(x)\} = \exp(\beta_0).\exp(\beta'\mu_1 + \frac{1}{2}\beta'\Sigma\beta),$$

from which it follows, using equation (8.14), that

$$\exp(\beta_0) = E\{\Omega(x)\}\exp(-\beta'\mu_1 - \frac{1}{2}\beta'\Sigma\beta)$$
$$= E\{\Omega(x)\}\exp\{-\frac{1}{2}\beta'(\mu_1 + \mu_2)\}.$$

Substituting into equation (8.13) then produces

$$\Omega(x) = E\{\Omega(x)\}\exp\{\beta'(x - \tfrac{1}{2}(\mu_1 + \mu_2)\}. \tag{8.16}$$

Equations (8.14) and (8.16) can form the basis of a method for estimating the selection function $\Omega(x)$. If random samples are taken from the two populations to provide sample mean vectors $\hat{\mu}_1$ and $\hat{\mu}_2$, and a sample pooled covariance matrix $\hat{\Sigma}$, then the vector β can be estimated by

$$\hat{\beta} = \hat{\Sigma}^{-1}(\hat{\mu}_2 - \hat{\mu}_1). \tag{8.17}$$

and the selection function by

$$\hat{\Omega}(x) = E\{\hat{\Omega}(x)\}\exp\{\hat{\beta}'(x - 1/2(\hat{\mu}_1 + \hat{\mu}_2)\}. \tag{8.18}$$

Obviously, if $E\{\Omega(x)\}$ is not known then it will only be possible to estimate $\Omega(x)$ multiplied by an arbitrary multiplicative scaling.

An approximation to the covariance matrix for the estimators $\hat{\beta}_1$ to $\hat{\beta}_p$ (which assumes that $\hat{\Sigma}$ is error free) is $\hat{\Sigma}^{-1}(1/n_1 + 1/n_2)$, where n_i is the size of sample i (Manly, 1985, p. 68). This will tend to be an underestimate, and bootstrapping is probably a better method for determining the covariance matrix.

Equations (8.17) and (8.18) can be used with any two multivariate normal distributions without the requirement that one of the populations is obtained from the other one by some selection process if $\Omega(x)$ is interpreted as the selection function that is required to change the first population into the second population. More generally, $\Omega(x)$ can be thought of as the function which says how many individuals there are in the second population with $X = x$ for each individual with these measurements in the first population. From this point of view, the function does not necessarily have to produce probabilities. In fact, there is no reason why values greater than one should be prohibited.

As with logistic regression, there are three situations that need to be considered in the context of estimating a resource selection probability function from two samples of resource units, corresponding to the samples being (a) a sample of available resource units and a sample of used resource units, (b) a sample of available resource units and a sample of unused resource units, and (c) a sample of unused resource units and a sample of used resource units. These three cases will now be considered in turn.

8.2.1 Samples of available and used resource units

Equations (8.17) and (8.18) apply immediately to the estimation of a resource selection probability function from a sample of available units and a sample of used units. Sample 1, with the mean vector $\hat{\mu}_1$, should be the available sample, and sample 2, with mean vector $\hat{\mu}_2$, should be the used sample. The selection function $\Omega(x)$ then has the same form as the resource selection probability function $w^*(x,t)$ of equation (7.5) when the selection time is t = 1. This therefore corresponds to the model for a small fraction of units used in Chapter 7. The quantity $E\{\Omega(x)\}$ in equation (8.18) is the proportion ρ of units used at the time that the second sample is taken.

8.2.2 Samples of available and unused resource units

If sample 1 for equations (8.17) and (8.18) is of available resource units, and sample 2 is of unused resource units then the function $\Omega(x)$ gives the probability that an individual with $X = x$ remains unused. Hence the resource selection probability function, $w^*(x,t) = 1 - \Omega(x)$ has the same form as equation (7.9) with $t = 1$. Thus in this situation the resource selection probability function can be estimated by

$$\hat{w}^*(x) = 1 - E\{\Omega(x)\}\exp\{\hat{\beta}'(x - 1/2(\hat{\mu}_1 + \hat{\mu}_2)\}.$$

Note that in this case $E\{\Omega(x)\}$ is the proportion of the available resource units that are unused at time t, $1 - \rho$, rather than the proportion used.

8.2.3 Samples of unused and used resource units

The situation is more complicated when a resource selection probability is to be estimated from a sample of unused resource units and a sample of used resource units. In this case, for every unit with $X = x$ in the available population there are $1 - w^*(x)$ units in the unused population and $w^*(x)$ units in the used population. Hence, for each unit with $X = x$ in the unused population there are $w^*(x)/\{1 - w^*(x)\}$ units in the used population. In other words, the distribution of X in the used population can be obtained from the distribution in the unused population by selecting individuals with $X = x$ with a probability proportional to $w^*(x)/\{1 - w^*(x)\}$.

What this implies is that if equations (8.17) and (8.18) are used, with sample 1 being unused units and sample 2 being used units, then $\Omega(x)$ corresponds to $w^*(x)/\{1 - w^*(x)\}$, and $E\{\Omega(x)\}$ must be interpreted as the average value of $\Omega(x)$ for the units in the unused sample. This means that

$$E\{\Omega(x)\}\exp\{\beta'(x - \tfrac{1}{2}(\mu_1 + \mu_2)\} = w^*(x)/\{1 - w^*(x)\}$$

so that the resource selection probability function is being assumed to have the form

$$w^*(x) = E\{\Omega(x)\}\exp\{\beta'(x - \tfrac{1}{2}(\mu_1 + \mu_2)\}/[1 + E\{\Omega(x)\}\exp\{\beta'(x - \tfrac{1}{2}(\mu_1 + \mu_2)\}].$$

Here $E\{\Omega(x)\}$ is the expected number of units in the used population for each unit in the unused population. Hence, if a proportion ρ of available units are used then $E\{\Omega(x)\} = \rho/(1-\rho)$. Substituting into the last equation then gives

$$w^*(x) = \frac{\{\rho/(1-\rho)\}\exp\{\beta'(x - \tfrac{1}{2}(\mu_1 + \mu_2)\}}{1 + \{\rho/(1-\rho)\}\exp\{\beta'(x - \tfrac{1}{2}(\mu_1 + \mu_2)\}}$$

which can be estimated by

$$\hat{w}^*(x) = \frac{\{\rho/(1-\rho)\}\exp\{\hat{\beta}\,'(x - \frac{1}{2}(\hat{\mu}_1 + \hat{\mu}_2)\}}{1 + \{\rho/(1-\rho)\}\exp\{\hat{\beta}\,'(x - \frac{1}{2}(\hat{\mu}_1 + \hat{\mu}_2)\}}.$$

Thus, to estimate the resource selection probability function from a sample of unused and a sample of used resource units, the β values should be estimated from equation (8.17), and substituted into the last equation, together with the sample means and the proportion of used units.

If ρ is not known then the estimation of the resource selection probability function, or even this function multiplied by an unknown scaling factor, becomes impossible. However, there are two ways to proceed that may be considered reasonable. First, setting $\rho = \frac{1}{2}$ (or any other arbitrary value) will give an estimated function that is monotonically related to the true resource selection probability function. Thus using this function will enable resource units to be ranked in order of their probability of use, so that it can be regarded as a type of selectivity index. Second, if ρ is small then $w^*(x)$ can be approximated by

$$\hat{w}^*(x) = \rho.\exp[\hat{\beta}\,'\{x - \frac{1}{2}(\hat{\mu}_1 + \hat{\mu}_2\,)\}],$$

so that an estimated resource selection function is

$$\hat{w}(x) = \exp(\hat{\beta}\,'x).$$

In effect this is equivalent to treating the sample of unused resource units as a sample of available resource units.

8.2.4 Example 8.2 Nest selection by fernbirds reconsidered

The linear discriminant function method can be used as an alternative to logistic regression to estimate a resource selection function from Harris' (1986) fernbird data in Table 2.4. Since there is a sample of available resource units (random points in the study region) and a sample of used resource units (nest sites), equations (8.17) and (8.18) can be applied directly.

The sample mean vector and covariance matrix for the 25 available nest sites (with the variables in the order X_1 = canopy height, X_2 = distance to edge and X_3 = perimeter of clump) are found to be

$$\hat{\mu}_1 = \begin{bmatrix} 0.4884 \\ 12.6200 \\ 2.9336 \end{bmatrix} \text{ and } \hat{\Sigma}_1 = \begin{bmatrix} 0.0216 & 0.1898 & -0.0007 \\ 0.1898 & 19.1308 & 1.5550 \\ -0.0007 & 1.5550 & 0.7128 \end{bmatrix}.$$

For the 24 nest sites the corresponding quantities are

$$\hat{\mu}_2 = \begin{bmatrix} 0.7325 \\ 16.2500 \\ 4.0367 \end{bmatrix} \text{ and } \hat{\Sigma}_2 = \begin{bmatrix} 0.0773 & -0.1613 & 0.1352 \\ -0.1613 & 11.6087 & -0.4115 \\ 0.1352 & -0.4115 & 1.8759 \end{bmatrix}.$$

The pooled sample covariance matrix is therefore

$$\hat{\Sigma} = \begin{bmatrix} 0.0489 & 0.0180 & 0.0568 \\ 0.0180 & 15.4498 & 0.5926 \\ 0.0658 & 0.5926 & 1.2820 \end{bmatrix},$$

with the inverse

$$\hat{\Sigma}^{-1} = \begin{bmatrix} 21.9899 & 0.0181 & -1.1378 \\ 0.0181 & 0.0659 & -0.0314 \\ -1.1378 & -0.0314 & 0.8530 \end{bmatrix}.$$

Substituting into equation (8.17) now produces the estimates

$$\hat{\beta} = \begin{bmatrix} \hat{\beta}_1 \\ \hat{\beta}_2 \\ \hat{\beta}_3 \end{bmatrix} = \begin{bmatrix} 4.1780 \\ 0.2090 \\ 0.5492 \end{bmatrix}.$$

The proportion of available sites that were used by the fernbirds, ρ, is not known, although it is obviously very small. Therefore, only a resource selection

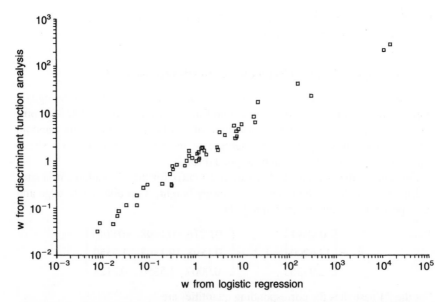

Figure 8.1 Comparison of values from resource selection functions estimated for nest site selection by fernbirds. The horizontal axis gives values obtained from a function estimated by logistic regression and the vertical axis gives values obtained from a function estimated using linear discriminant function equations. Logarithmic scales are used to accommodate the large range of values for the 49 sampled sites.

function that is proportional to the probability of selection can be estimated. Ignoring constant terms in the exponential argument this is

$$\hat{w}(x) = \exp\{4.18(\text{CANOPY}) + 0.21(\text{EDGE}) + 0.55(\text{PERIM})\}. \quad (8.19)$$

It is interesting to compare this estimated function with the function

$$\hat{w}(x) = \exp\{7.79(\text{CANOPY}) + 0.21(\text{EDGE}) + 0.88(\text{PERIM})\} \quad (8.20)$$

that was found in example 7.1 using logistic regression. On the face of it, these two functions agree only as far as the coefficient of the distance to edge is concerned. However, if the values obtained from equation (8.19) are plotted against the values obtained from equation (8.20) for all the 49 sampled resource units, as shown in Figure 8.1, then it is found that they are almost linearly related on a logarithmic scale, although there is far more variation in the estimates obtained from equation (8.20) than there is for those obtained from equation (8.19).

8.3 QUADRATIC DISCRIMINANT FUNCTION ANALYSIS

The theory developed in the previous section for the estimation of a resource selection function in terms of Fisher's linear discriminant function can be extended to allow selection functions of the form

$$w(x) = \exp\{(\beta'x + x'\Omega x)\},$$

where Ω is a symmetric p by p matrix, for describing how units must be selected from one population in order to produce the distribution of X in a second population. This means that the argument of the exponential function involves squares and products of the X variables as well as linear terms, which implies that the covariance matrices for the two populations can be different as well as the mean vectors. Using equations provided by Manly (1985, p. 66), it can be shown that $\log\{w(x)\}$ is in this case the usual quadratic discriminant function when the two populations have multivariate normal distributions for the X variables.

This approach for estimating a resource selection probability function will not be pursued further here on the grounds that if it is considered that this function should include squares and products of the X variables then it is more straightforward to include these terms in a log-linear model or logistic regression formulation.

8.4 DISCUSSION

In this chapter it has been shown that logistic regression can be used as an alternative to the log-linear modelling methods that were discussed in Chapter 7 when there are two samples of resource selection units being compared. There are three different possible cases, depending on whether the samples are of available, used or unused resource units.

Discriminant function analysis can also be used in these two sample situations if the variables that are measured on resource units have multivariate normal distributions in the two populations sampled. This gives a more efficient method of estimation when the assumptions are valid (Efron, 1975). However, distributions are typically not normal with resource selection studies so that in practice discriminant function analysis has limited value.

Exercise

Consider the data given in Table 7.2 for selection of *Daphnia publicaria* by yellow perch fry on 1 July, 1969. Note that this is a situation where there is a sample of available resource units and a sample of used resource units. Use logistic regression to estimate a resource selection function of the form $w(x) = \exp(\beta_1 X + \beta_2 X^2 + \beta_3 X^3)$, where X denotes the length of the *Daphnia* in mm. Verify that the estimated function is the same as that obtained in example 7.2, and use chi-squared tests to compare the fit of the cubic model with quadratic and linear models in X. Compare the estimated linear function with the estimate obtained using linear discriminant function methods.

9 General log-linear modelling

In this chapter the uses of log-linear modelling that have been discussed in Chapter 7 are extended to cover situations where resource selection can be related to factors such as the individual animals involved, or the time of day.

9.1 INTRODUCTION

The use of log-linear modelling has been discussed in Chapter 7 as a means of estimating a resource selection function using samples of used and available resource units, or as a means for estimating the relative 'survival' rates of different types of resource unit when only samples of unused units are available. However, as noted by Heisey (1985), the uses of log-linear modelling are by no means restricted to the situations that have been discussed in Chapter 7. In fact this approach to data analysis has a great deal of potential for modelling changes in selection related to factors such as the individual animal concerned, the time of day, the season of the year, etc.

With the general log-linear model it is assumed that the data available consists of I counts Y_1 to Y_I, where the ith count has a Poisson distribution that is independent of all the other counts, for which the mean value can be expressed as

$$\mu_i = \exp(\beta_0 + \beta_1 x_{i1} + \ldots + \beta_p x_{ip}), \tag{9.1}$$

where β_0 to β_p are constants to be estimated, and x_{i1} to x_{ip} are known values for the variables X_1 to X_p. The small fraction used model and the unused units only model of Chapter 7 can both be formulated within this framework providing that the X variables are defined in an appropriate manner.

Sometimes it is known that μ_i should be proportional to some known base rate, B_i, so that equation (9.1) is better written as

$$\mu_i = B_i \exp(\beta_0 + \beta_1 x_{i1} + \ldots + \beta_p x_{ip}). \tag{9.2}$$

For example, suppose that the data count Y_i is the observed number of times that a certain animal is seen in the ith type of habitat, where it is known that 60% of the habitat available to the animal is of this type. Then it is appropriate to set $B_i = 0.60$ so that if there is no selection then the model $\mu_i = B_i \exp(\beta_0)$ should fit the data, with the term $\exp(\beta_0)$ allowing for the total number of observations made on the animal.

9.2   FITTING MODELS TO DATA

Fitting a log-linear model requires that a suitable computer program is available. There are many such programs available, and there is no need to discuss the use of these here, other than to mention that in many cases these programs will construct sets of X variables to allow for the effects of categorical variables. For example, consider the data in Table 6.3 on the use by antelope of study plots with different distances from water and different aspects. Here the four aspects (East/Northeast, South/Southeast, West/Southwest and North /Northwest) are allowed for using the three 0–1 variables X_2 to X_4. Some computer programs will set up variables of this type automatically if the data input indicates that the data frequencies are classified by a factor at four levels. Note, however, there are alternative ways of defining the X variables for categorical variables, so that some computer programs would produce three X variables with different values from those shown in Table 6.3.

The goodness-of-fit of a log-linear model can be measured by the log-likelihood chi-squared statistic

$$X_L{}^2 = 2 \sum_{i=1}^{I} Y_i \log_e(Y_i/\hat{\mu}_i),$$

where $\hat{\mu}_i$ is the expected value of the ith data frequency according to the fitted model. There are $I - p - 1$ degrees of freedom. If this statistic is significantly large in comparison with the chi-squared distribution then the model is a poor fit to the data. As usual, most of the estimated expected values $\hat{\mu}_i$ should be five or more for the test to be reliable. As discussed in section 3.3, differences between the chi-squared statistics for different models can be tested against the chi-squared distribution to assess the relative goodness-of-fit of those models.

Log-linear models are so versatile that is not possible to cover all the situations where resource selection data can be analysed using these models. However, the following two examples illustrate the type of approach that can be used.

Example 9.1 Selection of forest canopy cover by elk

Consider Marcum and Loftsgaarden's (1980) data on the selection of forest over-story canopy cover by elk (*Cervus elaphus*). These data, which consist of counts in the four canopy over-story classes 0%, 1–25%, 26–75% and 76–100% for a sample of 200 points on a map of the study area, and an independent sample of 325 points selected by the population of elk, are given in Table 4.6. They have been discussed already in example 4.3 in the context of estimating selection ratios.

Since only a minute fraction of possible 'points' in the study region were sampled to determine the availability of the different canopy classes, and only a minute fraction of the possible 'points' were used, this study fits within what was called the small proportion used situation in Chapter 7. The data can therefore be analysed using the method proposed in Chapter 7, which amounts to fitting the log-linear model of equations (7.6) and (7.7) to the data. The only complication is the need to define appropriate X variables for equation (7.6) to allow for the four different classes of canopy cover. However, this just involves defining $X_1 = 1$ for a count on the 0% canopy class, or otherwise 0, $X_2 = 1$ for an observation on the 1–25% canopy class, or otherwise 0, and $X_3 = 1$ for an observation on the 26–75% canopy class, or otherwise 0. Thus three 0–1 X variables can be used to account for four classes of observations, in exactly the same way as has been done with several previous examples.

Another way of analysing the data, which gives essentially the same result as the method described in Chapter 7, involves setting up a log-linear model from first principles. Thus consider the data that are shown in Table 9.1. Here the first column gives the observed sample counts, while columns two to eight give the following X variables: $X_1 = 0$ for a count of available points and 1 for a count of used points; $X_2 = 1$ for the 0% canopy class, otherwise 0; $X_3 = 1$ for the 1–25% canopy class, otherwise 0; $X_4 = 1$ for the 26–75% canopy class, otherwise 0; X_5 = the product of X_1 and X_2; X_6 = the product of X_1 and X_3; and X_7 = the product of X_1 and X_4. Here the definitions of X_2 to X_7 are such that the availability of the 76–100% canopy class is being regarded as the 'standard' level of availability, and the selection for or against the 75–100% canopy class is taken as the 'standard' amount of selection.

Table 9.1 Data counts and X variables for a log-linear model for Marcum and Loftsgaarden's example on the selection of forest over-story canopy by elk

Data count	X variables						
Y	X_1	X_2	X_3	X_4	X_5	X_6	X_7
15	0	1	0	0	0	0	0
61	0	0	1	0	0	0	0
84	0	0	0	1	0	0	0
40	0	0	0	0	0	0	0
3	1	1	0	0	1	0	0
90	1	0	1	0	0	1	0
181	1	0	0	1	0	0	1
51	1	0	0	0	0	0	0

Using these variables, consider the log-linear model for which the expected value of the ith frequency is

$$\mu_i = \exp(\beta_0 + \beta_1 x_{i1} + \ldots + \beta_7 x_{i7}). \tag{9.3}$$

Here X_1 allows for the fact that the sample of used 'points' is not the same size as the sample of available 'points', X_2 to X_4 allow for the availability of the 0%, 1–25% and 26–75% canopy classes to be different from the availability of the 76–100% class, and X_5 to X_7 allow for the counts in the first three canopy classes to vary between the sample of available 'points' and the sample of used 'points'. Thus this is a log-linear model for the data which allows for the possibility of resource selection through the inclusion of variables X_5 to X_7.

If the 'no selection' model is fitted, using variables X_1 to X_4 only, then a chi-squared goodness-of-fit value of $X_{L0}^2 = 21.96$ with three degrees of freedom is obtained. This is clearly a very poor fit.

If the model is expanded to include the variables X_5 to X_8 then there are eight data counts and eight parameters in equation (9.3), which means that the log-linear model must fit the data exactly. This means that for this model allowing for selection the chi-squared goodness-of-fit statistic is $X_{L1}^2 = 0$, with zero degrees of freedom. The difference $X_{L0}^2 - X_{L1}^2 = 21.96 - 0 = 21.96$, with $3 - 0 = 3$ degrees of freedom, is then a measure of the improvement in the fit of the model due to allowing for habitat selection. Since this is very significantly large, there is very strong evidence of selection. It is interesting to note that this test for selection based on the comparison of the fit of log-linear models with and without an allowance for selection is exactly the same as the test using equation (4.19) for whether the sample of used resource units comes from the same population as the sample of available units.

Since the model allowing for selection must fit the data exactly, one way to determine estimates of the parameters β_0 to β_7 involves simply equating each of the observed counts to its expected value and solving the resulting eight equations. Alternatively, one of the standard iterative methods for estimating the parameters of log-linear models can be used, where this has the advantage of also producing standard errors for the estimated parameters.

Table 9.2 Estimated coefficients for the log-linear model (9.3) with standard errors and ratios of estimates to their standard errors

	Estimate	Standard error	Ratio
β_0, constant	3.690	–	–
β_1, used/unused indicator	0.243	0.211	1.15
β_2, 0% canopy class	−0.983	0.303	−3.25
β_3, 1–25% canopy class	0.421	0.203	2.07
β_4, 26–75% canopy class	0.741	0.192	3.86
β_5, use of 0% canopy class	−1.854	0.667	−2.78
β_6, use of 1–25% canopy class	0.146	0.268	0.54
β_7, use of 26–75% canopy class	0.525	0.249	2.11

Estimates with their standard errors are shown in Table 9.2. Since β_5 is more than two standard errors below zero, it appears that there is selection against the 0% canopy class, in comparison to the 76–100% class. On the other hand, since β_7 is more than two standard errors above zero it appears that there is selection in favour of the 26–75% canopy class, in comparison to the 76–100% class.

The parameters of the log-linear model can be related to the selection ratios that have been discussed in Chapter 4. In that chapter the selection ratio for type i resource units was defined to be the ratio of the proportion of used units to the proportion of available units of that type, which is proportional to the ratio of the number of used type i units to the number of available type i units. In the context of the log-linear model this implies that the selection ratio for the 0% canopy cover is proportional to

$$\mu_5/\mu_1 = \exp(\beta_1 + \beta_2 + \beta_5)/\exp(\beta_2) = \exp(\beta_1 + \beta_5),$$

taking into account the zero and one X values shown in Table 9.1. In a similar way it can be argued that the selection ratios for the 1–25%, 26–75% and 76–100% canopy classes are proportional to

$$\mu_6/\mu_2 = \exp(\beta_1 + \beta_6),$$
$$\mu_7/\mu_3 = \exp(\beta_1 + \beta_7)$$

and

$$\mu_8/\mu_4 = \exp(\beta_1),$$

respectively.

The implication of these results is that the estimated selection ratios $\hat{w}_1$ to $\hat{w}_4$ are proportional to $\exp(\hat{\beta}_5)$, $\exp(\hat{\beta}_6)$, $\exp(\hat{\beta}_7)$ and $\exp(0) = 1$, respectively, where the estimated β values are the ones produced by the log-linear model. For the data being considered, the β estimates are as shown in Table 9.2, so that the estimated selection ratios are proportional to $\exp(-1.854) = 0.157$, $\exp(0.146) = 1.157$, $\exp(0.525) = 1.690$ and 1, respectively. These are then the selection ratios shown in Table 4.6 multiplied by 0.785.

In summary, what this means is that the selection ratios that were estimated using equation (4.22) in example 4.3 are proportional to exponential functions of the parameters of the log-linear model that relate to selection. A similar result will apply whenever a log-linear model is fitted to data from a sample of available resource units and a sample of used resource units.

Of course, the calculations used in Chapter 4 are much more straightforward to use than a log-linear model, which means that in practice we would not recommend the use of a log-linear model for a simple situation like that of the present example. We stress, therefore, that our only reason for presenting this example was to illustrate the relationship between log-linear modelling and some of the methods that have been discussed in Chapter 4.

9.1.2 Example 9.2 Habitat selection by white-tailed Deer

A more useful application of log-linear modelling is described by Heisey (1985), using data reported by Nelson (1979) from a radio tracking study of habitat use by white-tailed deer (*Odoncoileus virginianus*). Here there are four habitat types, and relocations were observed for two deer at two times of the day. The counts of relocations in different habitats and the proportional availability of different habitats are shown in Table 9.3.

Table 9.3 Habitat use (HU) of two white-tailed deer in four types of habitat, with the proportional availability of those habitats (HA)

| Habitat | Midday | | | | Morning and evening | | | |
| | Deer 68 | | Deer 342 | | Deer 68 | | Deer 342 | |
	HU	HA	HU	HA	HU	HA	HU	HA
Aspen	18	0.66	29	0.65	43	0.66	46	0.65
Clearcut	2	0.20	1	0.13	33	0.20	29	0.13
Plantation	0	0.09	4	0.13	5	0.09	4	0.13
Spruce	0	0.05	0	0.09	0	0.05	2	0.09

In the terminology of Chapter 1, this is an example of a Design III study with sampling protocol A with availability censused and use sampled for each of two animals. Furthermore, since the resources being studied are defined by several categories, it is similar to the situation that has been discussed in section 4.14. There is, however, the complication that used resources were sampled at two times of day.

One method for analysing the data that might be considered involves using equation (4.40) to estimate selection ratios first using the midday results, and then separately for the morning and evening results. Equation (4.3) could then be used to estimate variances as explained in section 4.14. However, the problem with this approach is that the variance estimates would be rather unreliable since they would be based on differences between only two animals. Hence tests for selection and for differences between selection ratios would also be unreliable.

A log-linear model analysis overcomes these problems providing that it can be assumed that the counts in Table 9.3 are values from independent Poisson distributions. This will be reasonable providing that the individual observations on deer locations were far enough apart in time to be independent. We assume that this was the case.

Heissey used the computer program GLIM (McCullagh and Nelder, 1989) to analyse the data. This program automatically constructs X variables for the four habitat types, two deer, and two times of day. However, these variables can be set up easily enough for use in a program that does not have this facility, as shown in Table 9.4. It will be seen from this table that base rates and 12 variables are needed to account for selection related to the time of day and the deer.

The first column of Table 9.4 shows the counts of habitat use by the two deer, the second column gives the proportional availabilities of habitats, where these can be thought of as base rates B_i as in equation (9.2), variables X_1 to X_3 allow for sample counts to vary with the habitat in addition to the variation that is expected from the base rates, X_4 allows sample counts to vary with the time of day, X_5 allows sample counts to vary with the deer, X_6 to X_8 allow the effects of different habitats to vary with the time of the observations, X_9 to X_{11} allow the effects of different habitats to vary with the deer, and X_{12} allows the deer effect to vary with the time of day.

Table 9.4 Data for fitting log-linear models to account for the frequencies with which different types of habitat are used by white-tailed deer[*]

Sample count	Base rate	X_1	X_2	X_3	X_4	X_5	X_6	X_7	X_8	X_9	X_{10}	X_{11}	X_{12}
18	0.66	1	0	0	1	1	1	0	0	1	0	0	1
29	0.65	1	0	0	1	0	1	0	0	0	0	0	0
43	0.66	1	0	0	0	1	0	0	0	1	0	0	0
46	0.65	1	0	0	0	0	0	0	0	0	0	0	0
2	0.20	0	1	0	1	1	0	1	0	0	1	0	1
1	0.13	0	1	0	1	0	0	1	0	0	0	0	0
33	0.20	0	1	0	0	1	0	0	0	0	1	0	0
29	0.13	0	1	0	0	0	0	0	0	0	0	0	0
0	0.09	0	0	1	1	1	0	0	1	0	0	1	1
4	0.13	0	0	1	1	0	0	0	1	0	0	0	0
5	0.09	0	0	1	0	1	0	0	0	0	0	1	0
4	0.13	0	0	1	0	0	0	0	0	0	0	0	0
0	0.05	0	0	0	1	1	0	0	0	0	0	0	1
0	0.09	0	0	0	1	0	0	0	0	0	0	0	0
0	0.05	0	0	0	0	1	0	0	0	0	0	0	0
2	0.09	0	0	0	0	0	0	0	0	0	0	0	0

[*]The base rates are the proportions available of different habitats. The X variables are: $X_1 = 1$ for aspen, otherwise 0; $X_2 = 1$ for clearcut, otherwise 0; $X_3 = 1$ for plantation, otherwise 0; $X_4 = 1$ for midday, 0 for morning and afternoon; $X_5 = 1$ for deer 68, 0 for deer 342; $X_6 = X_1 X_4$; $X_7 = X_2 X_4$; $X_8 = X_3 X_4$; $X_9 = X_1 X_5$; $X_{10} = X_2 X_5$; $X_{11} = X_3 X_5$; and $X_{12} = X_4 X_5$.

On the basis of these variables, a 'no selection' model is one that includes X_4, X_5 and X_{12}. This allows the expected counts to depend on the deer and the time of day, with the deer effect possibly varying with the time of day, but says that habitat use is proportional to the availability. The chi-squared goodness-of-fit value is 78.26 with 12 degrees of freedom.

Adding X_1 to X_3 into the model allows for some selection to take place, where this is at the same level for both deer and both times of day. The resulting model has a goodness-of-fit statistic of 32.42 with nine degrees of freedom.

At this stage it is possible to expand the model to either allow selection to depend on the time of day (by adding X_6 to X_8 into the model) or to allow selection to depend on the deer (by adding X_9 to X_{11} into the model). It turns out that the goodness-of-fit is reduced considerably to 6.44 with six degrees of freedom if the first of these two options is taken, but reduced hardly at all to 30.88 with six degrees of freedom if the second option is chosen. The first option is therefore best.

The next stage in model building consists in allowing selection to depend on both the deer and the time of day. The model, which then includes all of the variables X_1 to X_{12}, has a chi-squared goodness-of-fit statistic of 4.41 with three degrees of freedom.

The only way that the model can be expanded at this stage is by adding X variables that allow the selection of habitat by deer to vary with the time of day. However, there would then be as many parameters as sample frequencies so that the model would be 'saturated' with parameters and fit the data exactly.

The model building process just described is summarised in the analysis of deviance shown in Table 9.5. A reasonable conclusion from this summary is that there was selection, and that this depended on the time of day but not on the deer.

Table 9.5   Analysis of deviance for the data on habitat selection by white-tailed deer

Model	X_L^2	df	Difference X_L^2	df
No selection of habitat	78.26*	12		
			45.84*	3
Constant selection on habitat	32.42*	9		
			25.98*	3
Selection varies with time	6.44	6		
			2.03	3
Selection varies with time and deer	4.41	3		
			4.41	3
Selection with the time effect varying with the deer	0.00	0		

*Significantly large at the 0.1% level.

10 Analysis of the amount of use

In this chapter consideration is given to situations where the amount of use is recorded for resource units, rather than just whether they are used or not. Three principal cases are recognized: (a) where the amount of use is a count (such as the number of animals present); (b) where the amount of use is measured (such as the biomass eaten); and (c) where the distribution of the amount of use is a mixture of zeros for unused units and a distribution of positive values for used units.

10.1 INTRODUCTION

Up to this point our discussion has been in terms of whether individual resource units are used or unused. However, in some cases data are also available on the amount of use received by the used units. This raises the question of how such data can be analysed.

We consider three situations in turn. First, it may be possible for a resource unit to be used by more than one animal. Then the amount of use can be recorded as the number of animals present, with zero indicating no use. Second, it may not be possible to know how many animals use a resource unit but the amount of the resource taken can be measured. Third, the distribution of the amount of use may consist of zeros for unused units and positive values for used units.

An example of the first case would be where quadrats are randomly placed in a study area and the number of animals of a certain species in each quadrat (U, say) is recorded together with variables X_1, X_2, ... X_P that measure the physical characteristics of the quadrats. It would then be interesting to relate the counts U to the X variables.

An example of the second case would be where quadrats are randomly placed in a study area and the biomass of vegetation eaten (Y, say) is recorded for each quadrat, along with the variables X_1 to X_P that measure the physical characteristics of the quadrat. Again, it would be interesting to relate the Y values to the X variables. We adopt the point of view that the important difference between this case and the first case is simply that the first case involves count data but the second case does not.

The example of the second case becomes an example of the third case if the biomass eaten on a quadrat is zero for quadrats that animals did not visit and greater than zero for the quadrats that were visited.

10.2 ANALYSIS OF COUNTS OF THE AMOUNT OF USE

It is useful to begin the consideration of the analysis of counts of the amount of use by discussing the proposition that all cases where the amount of use is counted can be considered as similar. The most obvious situation of interest is, of course, where the counts U recorded on the resource units are the numbers of individual animals found on the units. However, suppose instead that U is the number of 'signs' of use such as the number of tooth marks on trees. Then the 'signs' for a resource unit may have been made by more than one animal and the relationship between the number of 'signs' and the number of animals may be unknown. Still, establishing a relationship between the number of 'signs' and the characteristics of resource units as measured by variables X_1 to X_p may be a valuable contribution to the understanding of resource selection by the animal in question. Indeed, the number of animal 'signs' may be a more relevant measure of use than the number of animals if it is the effect of the animals on the resource units that is the main concern.

In some cases it is possible to relate the counts of the number of times that resource units are used to the X variables using standard statistical methods. In particular, a log-linear model may be suitable so that it can be assumed that the expected value of the amount of use of resource unit i is a Poisson random variable with mean value

$$\mu_i = \exp(\beta_0 + \beta_1 x_{i1} + \ldots + \beta_p x_{ip}), \tag{10.1}$$

where x_{ij} is the value of X_j for this unit.

Two problems are likely to occur with this approach. To begin with, it may be that the probability of including a resource unit in the sample varies according to the amount of use. This could be because used and unused resource units are sampled separately, or because there is a sampling bias caused by the fact that used units are more (or less) visible than unused units. The other problem is the natural tendency of many animal populations to be clustered, which means that the counts of the amount of use for different resource units will often show more variation than is expected from the Poisson distribution even when equation (10.1) gives the correct relationship between the expected amount of use and the X variables.

The first problem can only be overcome by explicitly recognizing that different resource units have different probabilities of being recorded. For example, suppose that the log-linear model for the counts of the amount of use of different resource units is correct, so that these counts have Poisson distributions with mean values given by equation (10.1), but the probability of recording information on an unused resource unit is $P_{\bar{u}}$ and the probability of recording information on a used resource unit is P_u. Then, if $P_{\bar{u}}$ and P_u are not equal the observed data will not follow the usual log-linear model. In fact, Bayes' theorem shows that the probability of observing zero use of a resource unit, conditional on that resource unit being sampled, is

$$\text{Prob}(U = 0 \mid \text{unit sampled}) = \frac{\text{Prob(unit sampled} \mid U = 0).\text{Prob}(U = 0)}{\text{Prob(unit sampled)}}$$

$$= \frac{P_{\bar{u}}\exp(-\mu)}{P_{\bar{u}}\exp(-\mu) + P_u\{1-\exp(-\mu)\}}$$

$$= \frac{\exp(-\mu)}{\exp(-\mu) + \{P_u/P_{\bar{u}}\}\{1-\exp(-\mu)\}} \qquad (10.2)$$

where μ is the unconditional expected count on the unit. Similarly, the probability of observing $U = u > 0$ is

$$\text{Prob}(U = 0 \mid \text{unit sampled}) = \frac{\text{Prob(unit sampled} \mid U = u).\text{Prob}(U = u)}{\text{Prob(unit sampled)}}$$

$$= \frac{P_u\exp(-\mu)\mu^u/u!}{P_{\bar{u}}\exp(-\mu) + P_u\{1-\exp(-\mu)\}}$$

$$= \frac{\{P_u/P_{\bar{u}}\}\exp(-\mu)\mu^u/u!}{\exp(-\mu) + \{P_u/P_{\bar{u}}\}\{1-\exp(-\mu)\}} \qquad (10.3)$$

Equations (10.2) and (10.3) define a two parameter modified Poisson distribution and together with equation (10.1) they give a modified log-linear model. Maximum likelihood estimates for the parameters of the model can be obtained by standard iterative methods, as outlined in the Appendix to this chapter.

The second problem mentioned above with modelling data on counts of the amount of use of resource units (more variation in counts than is expected with the Poisson distribution) can be handled in various ways. For example, it might be reasonable to assume that the amount of use for the ith unit is a negative binomial random variable with the mean value μ_i given by equation (10.1). The variance of the count is then greater than the Poisson variance (which is equal to μ_i). Alternatively, it can be assumed that a log-linear model is correct except that the variances of the counts are inflated by a constant heterogeneity factor. As discussed in section 2.4 and example 7.2, it is then valid to estimate parameters as for a log-linear model and simply adjust the standard errors of parameter estimators by multiplying by the square root of the estimated heterogeneity factor.

10.2.1 Example 10.1 Habitat selection by galaxiids

This example concerns a study carried out by McIntosh *et al.* (1992) using two stretches of the Shag River in North Otago, New Zealand. In one stretch the native fish *Galaxias vulgaris* was present but not the introduced brown trout (*Salmo trutta*). In the other stretch both species were present. The question addressed by the study was whether the selection of habitat by galaxiids is the same irrespective of whether trout are present or not.

Within the stretch of river without trout, three 15 m long sampling sites were chosen and each site was divided into approximately 210 quadrats with a size of 50 cm by 50 cm. Forty random quadrats were then chosen from each site and, for each of these quadrats, the number of galaxiids present was recorded together with the following 14 variables: X_1 = the width of the stream; X_2 = the distance to the nearest bank; X_3 = the proportion of bedrock; X_4 = the proportion of gravel; X_5 = the proportion of cobble; $X_6 = 1 - X_3 - X_4 - X_5$ = the proportion of boulder; X_7 = the mean surface area; X_8 = the maximum surface area; X_9 = the mean interstitial space; X_{10} = the maximum interstitial space; X_{11} = the left current velocity in the quadrat; X_{12} = the middle current velocity in the quadrat; X_{13} = the right current velocity in the quadrat; and X_{14} = the mean depth of the stream.

Three sites were also chosen in the stretch of water without trout. However, it was found that for these 'trout' sites the sampling scheme used for the 'no trout' sites was not satisfactory because of the low numbers of galaxiids present. To overcome this problem, the sampling area for each 'trout' site was extended to include about 40 m of river (approximately 560 quadrats) and a large random sample of quadrats was chosen and checked for the presence of galaxiids. A random subsample of 29 or 30 quadrats was then taken from the large sample, and augmented by a random subsample of between 10 and 13 quadrats randomly chosen from those containing galaxiids. It can be calculated that this sampling scheme gave a probability of approximately $P_{\bar{u}} = 0.049$ of sampling a quadrat without galaxiids and a probability of approximately $P_u = 0.331$ of sampling a quadrat with galaxiids at each of the 'trout' sites. For each of the subsampled quadrats the 14 variables indicated above were measured.

Since the 14 variables that characterize the quadrats are in some cases highly correlated and linearly related, it was decided to make a principal component analysis (Manly, 1986, Chapter 5) the first step in the treatment of the data. There were 237 sampled quadrats in all, and the principal component analysis was applied to these without regard to any differences between the six sites that were sampled. All variables were initially scaled to have unit variances. The outcome of the principal component analysis was that five principal components were found to have variances of more than one, which is commonly taken as a 'rule of thumb' for deciding how many components are important. Between them these five principal components accounted for 79.7% of the variation in the original data.

Table 10.1 shows the coefficients of the first five principal components after these principal components have been scaled to have a variance of unity over the 237 sampled quadrats. For example, the first principal component is

$$Z_1 = 0.22x_1 + 0.21x_2 - 0.17x_3 - 0.48x_4 - 0.17x_5 + 0.77x_6 + 0.97x_7 + 0.92x_8 + 0.97x_9 + 0.89x_{10} + 0.10x_{11} + 0.00x_{12} + 0.11x_{13} - 0.07x_{14},$$

Table 10.1   Coefficients of the first five principal components obtained from the data on all sampled quadrats[*]

X variable	Principal component				
	1	*2*	*3*	*4*	*5*
1. Width of stream	0.23	−0.40	−0.60	0.10	0.39
2. Distance to nearest bank	0.21	0.17	−0.71	0.16	0.41
3. Proportion of bedrock	−0.17	−0.21	0.54	0.61	0.39
4. Proportion of gravel	−0.48	−0.14	−0.50	0.31	−0.59
5. Proportion of cobble	−0.17	0.37	−0.01	−0.79	0.01
6. Proportion of boulder	0.77	0.08	0.12	−0.19	0.31
7. Mean surface area	0.97	−0.03	−0.03	−0.03	−0.06
8. Maximum surface area	0.92	−0.05	0.07	0.13	−0.25
9. Mean interstitial space	0.97	−0.04	−0.03	−0.01	−0.09
10. Maximum interstitial space	0.89	−0.05	0.07	0.14	−0.29
11. Left current velocity	0.10	0.72	−0.21	0.11	−0.04
12. Middle current velocity	0.00	0.81	0.00	0.11	0.04
13. Right current velocity	0.11	0.79	0.05	0.13	0.11
14. Mean depth of the stream	−0.07	0.69	0.04	0.34	−0.10

[*]These are coefficients for the standardized (mean zero and variance one) values of both the principal components and the original X variables.

where x_i is the value of the measured variable X_i after standardization to a mean of zero and a variance of one. From the coefficients in Table 10.1 it seems reasonable to identify the components as measuring the following properties of quadrats: Z_1, surface area and interstitial space; Z_2, velocity and depth of the stream; Z_3, width of stream; Z_4, bedrock and lack of cobble; and Z_5, gravel.

Following the principal component analysis, an attempt was made to relate the number of galaxiids found in quadrats to the values of the principal components. Because of the different sampling schemes at 'no trout' and 'trout' sites, it is convenient to analyse the data separately for the two cases. These separate analyses will now be described in turn, followed by a discussion of what can be learned from the whole study.

10.2.2 Analysis of results from the 'no trout' sites

A series of log-linear models were fitted to the data from the three 'no trout' sites to determine how well the numbers of galaxiids in quadrats can be accounted for using some or all of the first five principal components and dummy (0–1) variables allowing for site effects. The results obtained were as follows:

(A) The null model (model A) with the same expected number of galaxiids in each quadrat was fitted. This gave a goodness-of-fit statistic of 104.82 with 119 degrees of freedom.

(B) Next, model B included all five principal components and the dummy variables for site effects ($S_1 = 1$ for a quadrat from site 1, or otherwise 0; $S_2 = 1$ for a quadrat from site 2, or otherwise 0) was fitted. This gave a goodness-of-fit statistic of 86.19 with 112 degrees of freedom, which is significantly lower than the value from the null model (difference = 18.63 with 7 degrees of freedom, $p < 0.01$ by comparison with the chi-squared distribution).

(C) The third, fourth and fifth principal components were removed from the model one by one, without any significant increases in the goodness-of-fit statistic. This resulted in model C with a goodness-of-fit statistic of 91.29 with 115 degrees of freedom. The increase in the goodness-of-fit statistic over that of model B was not significant (difference = 5.10 with 3 degrees of freedom, $p > 0.1$).

(D) The second principal component was removed to produce model D. This resulted in a goodness-of-fit statistic of 94.45 with 116 degrees of freedom. The increase in the goodness-of-fit statistic over that of model C is not significant at the 5% level, although it is approaching significance (difference = 3.16 with 1 degree of freedom, $0.1 > p > 0.05$).

(E) The site effect variables S_1 and S_2 were removed to produce model E. This resulted in a significant increase in the goodness-of-fit statistic above the value for model D (difference = 8.63 with 2 degrees of freedom, $0.05 > p > 0.01$).

As will be seen below, there is clear evidence that the number of galaxiids was related to the first two principal components at the 'trout' sites. For this reason, it seems appropriate to accept model C as the best model for the 'no trout' sites even although the effect of the second principal component is not significant at the 5% level. According to this model,

$$\hat{\mu} = \exp(-1.80 + 0.45Z_1 + 0.30Z_2 - 1.80S_1 + 0.54S_2), \qquad (10.4)$$

so that the expected number of galaxiids in a quadrat was high for quadrats with large values for Z_1 (surface area and interstitial space) and large values for Z_2 (velocity and depth of the stream). Relative to site 3, there seem to have been low numbers of galaxiids at site 1 and slightly high numbers at site 2.

Table 10.2 shows the estimates that are obtained when a model relating galaxiid counts to Z_1 and Z_2 is fitted separately for each of the three sites. The consistency of the coefficients of these variables is reassuring since it indicates that the selection of habitat by galaxiids was much the same at each site. Furthermore, without going into details it can be mentioned that estimating separate coefficients for Z_1 and Z_2 for each of the three sites gives an insignificant reduction in the goodness-of-fit statistic of 2.37 with 4 degrees of freedom.

The observed frequencies of galaxiids ranged from 0 to 2 and the expected frequencies were all quite small. Therefore, it is questionable to compare the goodness-of-fit statistic of 91.29 with 115 degrees of freedom for model C with

Table 10.2   Results obtained from fitting log-linear models relating galaxiid counts at the 'no trout' sites to the first two principal components.

Parameter	Estimate	Standard error	Ratio
	Estimates for site 1		
Constant term	−1.24	0.32	–
Coefficient of Z_1	0.52	0.41	1.28
Z_2	0.25	0.45	0.56
	Estimates for site 2		
Constant term	−0.24	0.31	–
Coefficient of Z_1	0.72	0.34	2.10*
Z_2	0.17	0.24	0.70
	Estimates for site 3		
Constant term	−1.95	0.55	–
Coefficient of Z_1	0.27	0.30	0.90
Z_2	0.48	0.31	1.57
	Estimates for all sites		
Constant term	−1.80	0.40	–
Coefficient of Z_1	0.44	0.18	2.47*
Z_2	0.30	0.17	1.73
Coefficient of S_1	0.54	0.51	1.06
S_2	1.33	0.46	2.89*

*Significant at approximately the 5% level (outside the range ±1.96).

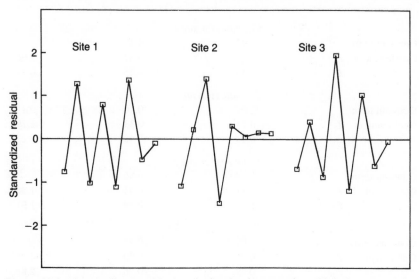

Figure 10.1   Standardized residuals calculated using equation (10.4) for sets of five quadrats from the 'no trout' sites

the percentage points of the chi-squared distribution in order to assess the absolute goodness-of-fit of the log-linear model. In fact, if this comparison is made then the fit seems to be unbelievably good since the probability of a goodness-of-fit statistic this low is about 0.002. A better idea of the adequacy of the model is obtained by considering Figure 10.1 which shows standardized residuals for groups of five quadrats. This figure was constructed by ordering the 40 quadrats within each site from the one with the smallest expected number of galaxiids to the one with the largest number of expected quadrats. The quadrats were then grouped into sets of five within the sites and the observed and expected number of galaxiids determined for each group. The standardized residuals (Observed − Expected)/√(Expected) were then calculated and plotted. These standardized residuals appear to have approximately standard normal distributions and the model therefore seems to be a good fit to the data.

10.2.3 Analysis of results from the 'trout' sites

The analysis carried out at the 'no trout' sites was repeated on the trout sites but with the modified log-linear model that is described by equations (10.1) to (10.3) and discussed further in the Appendix to this chapter. As noted above, the sampling probability for quadrats without galaxiids was approximately 0.049

Table 10.3   Results obtained from fitting modified log-linear models relating galaxiid counts to the first two principal components for data from the 'trout' sites

Parameter		Estimates	Standard error	Ratio
		Estimates for site 1		
Constant term		−2.93	0.47	–
Coefficient of	Z_1	0.94	0.32	2.96*
	Z_2	0.33	0.45	0.74
		Estimates for site 2		
Constant term		−2.69	0.40	–
Coefficient of	Z_1	0.75	0.33	2.29*
	Z_2	0.93	0.42	2.23*
		Estimates for site 3		
Constant term		−3.22	0.56	–
Coefficient of	Z_1	0.81	0.43	1.88
	Z_2	1.17	0.59	1.97*
		Estimates for All Sites		
Constant term		−2.91	0.26	–
Coefficient of	Z_1	0.86	0.20	4.21*
	Z_2	0.73	0.26	2.81*

*Significant at approximately the 5% level (outside the range ±1.96).

and the sampling probability for quadrats with galaxiids was approximately 0.331. Hence it was assumed for model fitting that $P_u/P_{\bar{u}} = 0.331/0.049 = 6.7$.

It was found by first fitting a model with effects for the first five principal components, and then removing nonsignificant terms one by one, that only the first two of the principal components seem to have important effects. With just these terms in the model it is estimated that the expected number of galaxiids in a quadrat is given by

$$\hat{\mu} = \exp(-2.91 + 0.86Z_1 + 0.73Z_2). \tag{10.5}$$

Table 10.3 shows that the coefficients of Z_1 and Z_2 are both quite large relative to their standard errors, and that quite similar estimates are obtained if the model is fitted separately to the data from the three sites. Furthermore, the improvement in fit that is obtained by estimating the parameters separately for each site is small and insignificant (difference in goodness-of-fit statistics = 3.67 with 6 degrees of freedom).

10.2.4 Analysis of combined results

The coefficients of Z_1 and Z_2 in equations (10.4) and (10.5) are quite similar. In fact, the two equations indicate virtually the same habitat selection by the galaxiids at the 'no trout' and 'trout' sites, as can be seen from Figure 10.2, which shows the logarithm of the values obtained from equation (10.5) plotted against the logarithms of the values obtained from equation (10.4) (with $S_1 = S_2 = 0$ to remove site effects). To be more precise, logarithms of the functions (10.4) and (10.5) were evaluated for the 120 sampled quadrats at the 'no trout' sites. This provided the 120 points shown in part (a) of the figure. The functions (10.4) and (10.5) were also evaluated for the 117 sampled quadrats at the 'trout' sites. This provided the 117 points shown in part (b) of the figure. Note that logarithms were plotted simply to avoid 'bunching' of the points for the relatively large number of quadrats with small estimated means.

The plotted values show an extremely high correlation (overall: n = 237, r = 0.992, p < 0.001). Hence, although there were far fewer galaxiids at the 'trout' sites than at the 'no trout' sites, the function estimated from the 'no trout' sites predicts very well the habitat use at the 'trout' sites, and the function estimated at the 'trout' sites predicts very well the habitat use at the 'no trout' sites.

A comparison of parts (a) and (b) of Figure 10.2 indicates that the distribution of the values of equations (10.4) and (10.5) are virtually the same for the quadrats from the 'no trout' sites and the quadrats from the 'trout' sites. Thus the 'no trout' and 'trout' sites seem to be inherently about as attractive to galaxiids. This raises the question of why there are so many fewer galaxiids at the 'trout' sites. The obvious answer is the presence of trout which could inhibit the galaxiid population either by predating on it or by competing for space. However, since the three 'no trout' sites and the three 'trout' sites are really pseudoreplicates (Hurlbert, 1984) taken from areas that were not chosen

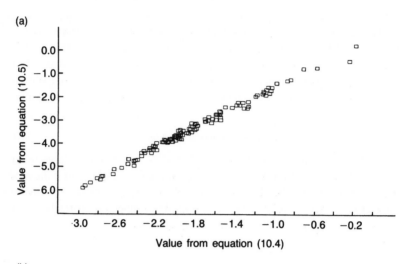

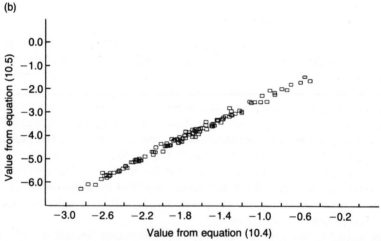

Figure 10.2 Plot of logarithms of mean numbers of galaxiids in quadrats as estimated from the data from 'trout' sites (equation (10.5)) against mean numbers estimated from the data from 'no trout' sites (equation (10.4)): (a) plot for the 120 quadrats from 'no trout' sites; (b) plot for the 117 quadrats from 'trout' sites.

at random, this may or may not be true. The galaxiid numbers may be related to some entirely different factor that has not been recognized in this study.

10.3 ANALYSIS OF CONTINUOUS MEASURES OF THE AMOUNT OF USE

Suppose now that the amount of use of a resource unit is a continuous variable Y with all resource units having the same probability of inclusion in the sample to be analysed, and with no complications from a substantial part of the population consisting of unused units with zero measured use. In that case it may well be that standard methods of statistical analysis will be sufficient to determine the relationship between the amount of use and variables measured on resource units. For example, the linear regression model

$$Y = \beta_0 + \beta_1 X_1 + \dots + \beta_p X_p + \epsilon$$

might be used to relate Y to the X variables that measure characteristics of the resource unit. Also, if the assumptions of the linear regression model are not valid then it may be possible to transform the data so that the model can be used. Alternatively, a non-linear regression might be considered.

There are obviously endless possibilities in this type of situation and it is difficult to make any further general comments.

10.4 MIXTURES OF ZEROS AND POSITIVE MEASURES OF THE AMOUNT OF USE

The final situation that we consider is where the distribution of the amount of use Y of a resource consists of 0 with probability p and values from a positive distribution with probability 1–p. Thus p is the probability that a resource unit is not used, so that Y is necessarily zero, and the continuous distribution is the distribution of Y conditional on use.

Models of this type have been considered by Lachenbruch (1976) for the situation where two samples need to be compared to see whether the probability p or the positive distribution differs in the populations that the samples are drawn from. Exponential and log-normal distributions were considered for the non-zero values in the samples, as well as the Poisson distribution with zero values omitted. Lachenbruch discussed maximum likelihood estimation of the parameters of the two part distribution and showed that the maximum likelihood estimate of p is the sample proportion of zeros and that the parameters of the non-zero distribution are exactly the same as is found by ignoring the zero data. Furthermore, the estimates of p and the parameters of the non-zero distribution are independent.

In the context of resource selection the situation is more complicated since the probability of a unit being used and the amount of use of a unit are likely to be functions of the characteristics of that unit as measured by the variables X_1 to X_p. However, it is still true that the estimation of the probability of use and the distribution of the amount of use can be considered separately.

To see this, suppose that the values of the X variables for the ith resource unit are $x_{i1}, x_{i2}, \dots, x_{ip}$, and that the probability of this unit being used is given by a function $p(x_{i1}, x_{i2}, \dots, x_{ip})$. Suppose also that if this unit is used then the probability density function of the amount of use Y_i of the unit is $f(y_i; x_{i1}, x_{i2}, \dots, x_{ip})$. Then if data are collected on n resource units, of which the first n_0 are unused, the likelihood function for the data takes the form

$$L = \prod_{i=1}^{n_0} \{1-p(x_{i1},x_{i2},\ldots,x_{ip})\} \prod_{i=n_0+1}^{n} p(x_{i1},x_{i2},\ldots,x_{ip})\ f(y_i;x_{i1},x_{i2},\ldots,x_{ip}).$$

Hence the log-likelihood function is

$$\log(L) = \sum_{i=1}^{n_0} \log\{1-p(x_{i1},x_{i2},\ldots,x_{ip})\} \sum_{i=n_0+1}^{n} \log\{p(x_{i1},x_{i2},\ldots,x_{ip})\}$$

$$+ \sum_{i=n_0+1}^{n} \log\{f(y_i;x_{i1},x_{i2},\ldots,x_{ip})\}.$$

It can be seen that this log-likelihood function is in two parts. The first two terms on the right-hand side relate to the probability of use, while the last term relates to the amount of use conditional on a unit being used. This implies that the maximum likelihood estimators of any parameters of the function p are not a function of the measured amounts of use y_i on the used units, and that the maximum likelihood estimators of any parameters of the function f are not a function of the X values for unused units. In practice, therefore, the two functions p and f can be estimated quite separately.

There is a reservation here that must be made. If the functions p and f have any shared parameters then, of course, maximum likelihood estimation must be carried out using all the data simultaneously taking this into account. For example it might be reasonable to assume that p and f are both functions of the same linear combination $\beta_0 + \beta_1 X_1 + \ldots + \beta_p X_p$ of the X variables, where this measures the desirability of a resource unit both in terms of whether it is used at all and, if so, how much it is used. In that case estimating p and f separately will not be satisfactory.

The methods used for separate estimation of p and f will obviously depend on the nature of the data. However, a reasonable approach to try would involve estimating the probability of use function with a logistic regression as discussed in Chapter 5, and then estimating the amount of use function using multiple regression.

APPENDIX: MAXIMUM LIKELIHOOD ESTIMATION OF A MODIFIED LOG-LINEAR MODEL

In this Appendix the maximum likelihood estimation of the parameters Θ, β_0, β_1, ..., β_p of the model specified by equations (10.1) to (10.3) are considered for the case where the data available consists of the values of variables X_1 to X_p for each of n_0 unused resource units, and for each of n_1 resource units used at least once. For the used resource units the number of times used must also be known.

In this situation the likelihood function (the probability of the observed data as a function of the unknown parameters) is given by

$$L = \prod_{i=1}^{n_0} 1/[1+\Theta\{\exp(\mu i)-1\}] \prod_{i=n_0+1}^{n_0+n_1} [\{P_u/P_{\bar{u}}\}\mu_i{}^{u_i}/u_i!]/[1+\{P_u/P_{\bar{u}}\}\{\exp(\mu_i)-1\}], \quad (A.1)$$

where the labelling of the resource units is such that the first n_0 are the unused ones and the last n_1 are the used ones, $\mu_i = \exp(\beta_0 + \beta_1 x_{i1} + \ldots + \beta_p x_{ip})$ is the expected number of times that the ith unit will be used, and u_i is the observed amount of use for the ith unit.

The likelihood function (A.1) is a function of $P_u/P_{\bar{u}}\}$, β_0, β_1, ..., β_p. Therefore, maximum likelihood estimates of some or all of these parameters can be obtained by maximizing L or its logarithm $\log_e(L)$. Equations for finding these estimates can be found in the usual way by equating the derivatives of $\log_e(L)$ to zero for each of the parameters. However, these equations do not have an explicit solution so that some numerical method for finding the estimates must be used. The Newton–Raphson method (Manly, 1985, p. 405) seems to work very well for this purpose, and was what was used for the calculations for the 'trout' site data of example 10.1.

Although there is no simple chi-squared test for the goodness-of-fit of this model (since the observed data will not have Poisson distributions), it is still possible to compare the goodness-of-fit of two models if one of the models is a special case of the other. Thus if $\log_e(L_1)$ is the maximized log-likelihood for the special case model, with p_1 estimated parameters, and $\log_e(L_2)$ is the maximized log-likelihood for the more general model, with $p_1 + p_2$ parameters (of which the first p_1 are the same as for the first model), then for large samples $-2\{\log_e(L_2) - \log_e(L_1)\}$ can be compared with the chi-squared distribution with p_2 degrees of freedom to see if the more general model is a significantly better fit than the special case model. In effect, this means that the value of $-2\log_e(L)$ can be treated as a measure of the goodness-of-fit of a model, although this cannot be compared directly with critical values of the chi-squared distribution.

11 The comparison of selection for different types of resource unit

This final chapter is concerned with how the selection for different types of resource can be compared when the amount of selection is estimated by using a logistic, proportional hazards or log-linear model.

11.1 VARIANCES FOR ESTIMATES AND THEIR DIFFERENCES

With log-linear models and many of the other models that have been considered in earlier chapters, the amount of selection for or against a particular type of resource unit, or the comparison between the selection for two types of resource units involves the consideration of exponential functions of estimated parameters. It is therefore useful to conclude this book with a review of statistical methods that can be used to assess the accuracy of estimates of exponential functions and their differences.

First, suppose that a logistic function for the probability that a resource unit with measurement x_1 to x_p has been estimated as described in Chapter 5. The estimated function then takes the form

$$\hat{w}^*(x) = \frac{\exp(\hat{\beta}_0 + \hat{\beta}_i x_1 + \ldots + \hat{\beta}_p x_p)}{1 + \exp(\hat{\beta}_0 + \hat{\beta}_1 x_1 + \ldots + \beta_0 x_p)}. \tag{11.1}$$

The variance of this function can be determined using the Taylor series method (Manly, 1985, p. 408) to be approximately

$$\text{var}\{\hat{w}^*(x)\} = w^*(x)^2 \{1 - w^*(x)\}^2 \sum_{i=0}^{p} \sum_{j=0}^{p} x_i x_j \text{cov}(\hat{\beta}_i, \hat{\beta}_j), \tag{11.2}$$

taking $x_0 = 1$. Here $\text{cov}(\hat{\beta}_i, \hat{\beta}_j)$ is the variance of $\hat{\beta}_i$ if $i = j$, or is otherwise the covariance between $\hat{\beta}_i$ and $\hat{\beta}_j$, where these variances and covariances should be available as part of output from the computer program used to fit the logistic function.

The Taylor series method also shows that if the difference between the probability of use for a resource unit with X values $x_1 = (x_{11}, \ldots, x_{1p})$ and the probability of use for a resource unit with $x_2 = (x_{21}, \ldots, x_{2p})$ is estimated by $\hat{w}^*(x_1) - \hat{w}^*(x_2)$, then this estimator has the approximate variance

$$\text{var}\{\hat{w}^*(x_1) - \hat{w}^*(x_2)\} = w^*(x_1)\{1 - w^*(x_1)\}w^*(x_2)\{1 - w^*(x_2)\}$$

$$.[\sum_{i=0}^{p} \sum_{j=0}^{p} x_{1i}x_{2j}\text{cov}(\hat{\beta}_i\hat{\beta}_j)], \tag{11.3}$$

taking $x_{10} = x_{20} = 1$.

In Chapter 6 the estimated resource selection probability function

$$\hat{w}^*(x) = 1 - \exp\{-\exp(\hat{\beta}_0 + \hat{\beta}_1 x_1 + \ldots + \hat{\beta}_p x_p)t\} \tag{11.4}$$

was used, where this is based on the proportional hazards survival function for the survival to time t. Here the Taylor series method gives the approximate variances

$$\text{var}\{w^*(x)\} = w^*(x)^2[\log_e\{1 - w^*(x)\}]^2 \sum_{i=0}^{p} \sum_{j=0}^{p} x_i x_j \text{cov}(\hat{\beta}_i, \hat{\beta}_j), \tag{11.5}$$

and

$$\text{var}\{\hat{w}^*(x_1) - \hat{w}^*(x_2)\} = w^*(x_1)\log_e\{1 - w^*(x_1)\}w^*(x_2)\log_e\{1 - w^*(x_2)\}$$

$$.[\sum_{i=0}^{p} \sum_{j=0}^{p} x_{1i}x_{2j}\text{cov}(\hat{\beta}_i, \hat{\beta}_j)], \tag{11.6}$$

taking $x_0 = 1$ in equation (11.5) and $x_{10} = x_{20} = 1$ in equation (11.6).

The last two equations can also be used if the function (11.4) is estimated from sample data, as described in Chapter 6. With all applications of the equations the variances and covariances of the β estimates will need to be determined from the output of the computer program used for estimation.

The other situation that needs to be considered is where the estimated function is of the form

$$\hat{f}(x) = \exp(\hat{\beta}_0 + \hat{\beta}_1 x_1 + \ldots + \hat{\beta}_p x_p). \tag{11.7}$$

Then the Taylor series method gives the approximation

$$\text{var}\{\hat{f}(x)\} = f(x)^2 \sum_{i=0}^{p} \sum_{j=0}^{p} x_i x_j \text{cov}(\hat{\beta}_i, \hat{\beta}_j), \tag{11.8}$$

and

$$\text{var}\{\hat{f}(x_1) - \hat{f}(x_2)\} = \hat{f}(x_1)\hat{f}(x_2) \sum_{i=0}^{p} \sum_{j=0}^{p} x_{1i}x_{2j}\text{cov}(\hat{\beta}_i, \hat{\beta}_j)], \tag{11.9}$$

taking $x_0 = 1$ in equation (11.7) and $x_{10} = x_{20} = 1$ in equations (11.8) and (11.9).

Functions of the form of equation (11.7) have been suggested for modelling both resource selection functions, and also for modelling relative survival rates. In both cases it may be desirable to compare the ratio of two function values rather than the difference. To this end it can be noted that

$$\hat{f}(x_1)/\hat{f}(x_2) = \exp\{\hat{\beta}_1(x_{11}-x_{21}) +...+ \hat{\beta}_p(x_{1p}-x_{2p})\},$$

so that equation (11.18) provides the result

$$\text{var}\{\hat{f}(x_1)/\hat{f}(x_2)\} = \{\hat{f}(x_1)/\hat{f}(x_2)\}^2$$

$$.[\sum_{i=1}^{p} \sum_{j=1}^{p} (x_{1i}-x_{2i})(x_{1j}-x_{2j})\text{cov}(\hat{\beta}_i,\hat{\beta}_j). \qquad (11.10)$$

11.1.1 Example 11.1 The logistic selection function for habitat selection by pronghorn

Example 5.1 was concerned with the selection of winter habitat by pronghorn in the Red Rim area of south-central Wyoming. The details of the analysis used will not be repeated here. It will merely be noted that the probability of a study plot being used (the resource selection probability function) was modelled by a logistic equation, and it was concluded that the function should include X variables allowing for the year involved, the distance from water, and the aspect of the study plot.

The estimated resource selection probability function is

$$w^* = \frac{\exp\{-0.613+0.410(\text{YEAR}) - 0.00037(\text{DW})+0.786(\text{E/NE})--0.059(\text{S/SE})+0.180(\text{W/SW})\}}{1 + \exp\{-0.613+0.410(\text{YEAR})-0.00037(\text{DW})+0.786(\text{E/NE})-0.059(\text{S/SE})+0.180(\text{W/SW})\}}, \qquad (11.11)$$

where YEAR is 0 for 1980–81 and 1 for 1981–82, DW is the distance to water in metres and E/NE, S/SE and W/SW are variables that are 1 for study plots with the aspect indicated, but are otherwise 0. The matrix of variances and covariances for the estimated constant term –0.613 and the coefficients of YEAR, DW, E/NE, S/SE and W/SW were output from the computer program used to carry out the estimation.

One application of equation (11.2) is to find confidence intervals for true probabilities of use. For example, consider just the East/Northeast study plots in 1980–81. For these plots, the variables YEAR, S/SE and W/SW are always zero, which means that the variances and covariances associated with these terms do not contribute to the sum on the right-hand side of equation (11.2). A further simplification is that the E/NE variable is always equal to one. The result is that the equation for the variance of w^* is fairly straightforward to apply.

Figure 11.1 shows the estimated resource selection probability function for study plots at different distances to water, with approximate 95% confidence limits that are the estimated probabilities plus and minus 1.96 estimated standard errors. It is apparent that the resource selection probability function is not well estimated in this case.

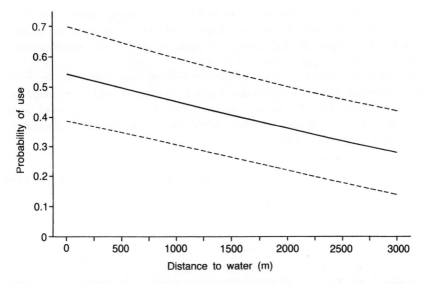

Figure 11.1 Probabilities of use by pronghorn for East/Northeast study plots in 1980/81, as a function of the distance to water. The estimated resource selection probability function $\hat{w}^*$ is the continuous line and the broken lines are approximate 95% confidence intervals given by $\hat{w}^* \pm 1.96\text{se}(\hat{w}^*)$.

11.1.2 Example 11.2 The relative survival of different types of corixid

Consider next the situation discussed in example 7.3 where samples of surviving corixids were used to estimate a relative survival function. As in the previous example, there is no need to repeat the details of the analysis given earlier. Instead, it will simply be noted that the probability of a corixid surviving t days was estimated to be proportional to

$$\hat{\phi}(x,t) = \exp\{(0.105x_1 + 0.0630x_2 - 0.156x_3 + 0.093x_4)t\}, \quad (11.12)$$

where $x_1 = 1$ for *Sigara venusta*, otherwise 0; $x_2 = 1$ for *S. praeusta*, otherwise 0; $x_3 = 1$ for light-coloured corixids, otherwise 0; and $x_4 = 1$ for medium-coloured corixids, otherwise 0.

As part of the fitting process, the covariance matrix for the four parameters in the equation was estimated to be

$$\begin{bmatrix} 0.00146 & 0.00140 & 0.0000622 & 0.0000315 \\ 0.00140 & 0.00252 & -0.0000735 & 0.0000108 \\ 0.0000622 & -0.0000735 & 0.00123 & 0.000248 \\ 0.0000315 & 0.0000108 & 0.000248 & 0.000301 \end{bmatrix}$$

Using the elements of this matrix it is possible to estimate the variance of the relative survival function for any corixid using equation (11.8), the variance of the difference between two values from the function using equation (11.9), or the variance of the ratio of two values from the function using equation (11.10).

Table 11.1 shows the estimates from equation (11.12) with standard errors determined as the square roots of variances obtained using equation (11.8), together with approximate 95% simultaneous Bonferroni confidence limits for the population relative survival rates. From these limits it can be concluded that medium-coloured S. venusta and medium-coloured S. distincta survived better than the 'standard' dark S. distincta, while light-coloured S. distincta did not survive as well as dark S. distincta.

Table 11.1 Estimated relative survival rates for corixids, together with estimated standard errors and approximate Bonferroni simultaneous 95% confidence limits[*]

Species	Colour	Estimated relative survival	Standard error	Confidence limits	
				Lower	Upper
S. venusta	Light	0.950	0.050	0.811	1.088
	Medium	1.218	0.052	1.075	1.362
	Dark	1.110	0.042	0.994	1.227
S. praeusta	Light	0.908	0.054	0.758	1.058
	Medium	1.165	0.062	0.994	1.336
	Dark	1.061	0.053	0.915	1.208
S. distincta	Light	0.855	0.030	0.773	0.938
	Medium	1.097	0.019	1.045	1.149
	Dark	1.000	0.000	–	–

[*]Since dark *Sigara distincta* are given a relative survival rate of exactly one, there are eight confidence intervals for differences from this value for other types of corixid. On this basis it is appropriate to set the confidence level for each parameter at $(100-5/8)\%$ = 99.4% in order to have a confidence of 95% that all limits will include the true parameter value. From standard normal tables this suggests taking the limits as estimate $\pm$ 2.75(standard error).

As an illustration of how to calculate the standard error for an estimated ratio, suppose that it is desirable to estimate the standard error for the ratio of the survival rate of light-coloured S. distincta to the survival rate of medium–coloured S. venusta. The estimate is then found to be

$$\frac{\exp\{0.105(0) + 0.060(0) - 0.156(1) + 0.093(0)\}}{\exp\{0.105(1) + 0.060(0) - 0.156(0) + 0.093(1)\}} = 0.702,$$

where the 0 and 1 values in parenthesis are the values of x_1 to x_4 for the two types of corixid. Using these values in equation (11.10), together with the appropriate variances and covariances from the covariance matrix then yields the following variance for this estimated ratio

$$0.702^2\{(-1)^2 \times 0.00146 + 1^2 \times 0.00123 + (-1)^2 \times 0.000301$$
$$+ 2 \times (-1)(1) \times 0.0000622 + 2(-1)(-1) \times 0.0000315$$
$$+ 2 \times (1)(-1) \times 0.000248\} = 0.001199.$$

The standard error associated with the estimate of 0.702 is therefore $\sqrt{0.001199} = 0.035$. Thus an approximate 95% confidence interval for the ratio of the survival probability of light *S. distincta* to medium *S. venusta* is $0.702 \pm 1.96 \times 0.035$, or from 0.634 to 0.770.

References

Alldredge, A.W., Deblinger, R.D. and Peterson, J. (1991). Birth and bedsite selection by pronghorns in a sagebrush steppe community. *Journal of Wildlife Management* **55**: 222–7.

Alldredge, J.R. and Ratti, J.T. (1986). Comparison of some statistical techniques for analysis of resource selection. *Journal of Wildlife Management* **50**: 157–65.

Alldredge, J.R. and Ratti, J.T. (1992). Further comparison of some statistical techniques for analysis of resource selection. *Journal of Wildlife Management* **56**: 1–9.

Arnett, E.B., Cook, J.G., Lindzey, F.G. and Irwin, L.L. (1989). Encampment River bighorn sheep study, June 1987 – December 1988 summary. Department of Zoology and Physiology, University of Wyoming, Laramie, Wyoming.

Bantock, C.R., Bayley, J.A. and Harvey, P.H. (1976). Simultaneous selective predation on two features of a mixed sibling species population. *Evolution* **29**: 636–49.

Beamsderfer, R.C. and Rieman, B.E. (1988). Size selectivity and bias in estimates of population statistics of small mouth bass, walleye, and northern squawfish in a Columbia River reservoir. *North American Journal of Fisheries Management* **8**: 505–10.

Belaud, A., Lim, P. and Sabaton, C. (1989). Probability-of-use curves applied to brown trout (*Salmo trutta Fario* L.) in rivers of southern France. *Regulated Rivers: Research and Management* **3**: 321–36.

Belovsky, G.E., Ritchie, M.E. and Moorehead, J. (1989). Foraging in complex environments: when prey availablity varies over time and space. *Theoretical Population Biology* **36**: 144–60.

BMDP (1988). *The SOLO Statistical System.* BMDP Statistical Software Inc., Los Angeles.

Bovee, K.D. (1981). *A User's Guide to the Instream Flow Incremental Methodology.* United States Fish and Wildlife Service Report FWS/OBS–78/07, Fort Collins, Colorado.

Bowyer, R.T. and Bleich, V.C. (1984). Effects of cattle grazing on selected habitats of southern mule deer. *California Fish and Game* **70**: 240–7.

Bryant, E.H. (1973). Habitat selection in a variable environment. *Journal of Theoretical Biology* **41**: 421–9.

Byers, C.R., Steinhorst, R.K. and Krausman, P.R. (1984). Clarification of a technique for analysis of utilization-availability data. *Journal of Wildlife Management* **48**: 1050–3.

Chesson, J. (1978). Measuring preference in selective predation. *Ecology* **59**: 211–15.

Cochran, W.G. (1977). *Sampling Techniques*, 2nd edn. Wiley, New York.

Cock, M.J.W. (1978). The assessment of preference. *Journal of Animal Ecology* **47**: 805–16.

Colgon, P.W. and Smith, J.T. (1985). Experimental analysis of food preference transitivity in Fish. *Biometrics* **41**: 227–36.

Cox, D.R. and Hinkley, D.V. (1974). *Theoretical Statistics*. Chapman and Hall, London.

Danell, K., Edenius, L. and Lundberg, P. (1991). Herbivory and tree stand composition: moose patch use in winter. *Ecology* **72**: 1350–7.

Dubuc, L.J., Krohn, W.B. and Owen, R.B. (1990). Predicting occurrence of river otters by habitat on Mount Desert Island, Maine. *Journal of Wildlife Management* **54**: 594–9.

Dunn, P.O. and Braun, C.E. (1986). Summer habitat use by adult female and juvenile sage grouse. *Journal of Wildlife Management* **50**: 228–35.

Edge, W.D., Marcum, C.L. and Olson–Edge, S.L. (1987). Summer habitat selection by elk in Western Montana: a multivariate approach. *Journal of Wildlife Management* **51**: 844–51.

Edwards, T.C. and Collopy, M.W. (1988). Nest tree preference of osprey in Northcentral Florida. *Journal of Wildlife Management* **52**: 103–7.

Efron, B. (1975). The efficiency of logistic regression compared to normal discriminant analysis. *Journal of the American Statistical Association* **70**: 892–8.

Ellis, J.E., Wiens, J.A., Rodell, C.F. and Anway, J.C. (1976). A conceptual model of diet selection as an ecosystem process. *Journal of Theoretical Biology* **60**: 93–108.

Emlen, J. M. (1966). The role of time and energy in food preference. *American Naturalist* **100**: 611–17.

Fagen, R. (1988). Population effects of habitat change: a quantitative assessment. *Journal of Wildlife Management* **52**: 41–6.

Feder, J.L., Chilcote, C.A. and Bush, G.L. (1990). The geographic pattern of genetic differentiation between host associated populations of *Rhagoletis pomonella* (Diptera: Tephritidae) in the eastern United States and Canada. *Evolution* **44**: 570–94.

Forsman, E.D., Meslow, E.C. and Wight, H.M. (1984). Distribution and biology of the spotted owl in Oregon. *Wildlife Monographs* **87**: 1–64.

Friedman, M. (1937). The use of ranks to avoid the assumption of normality implicit in the analysis of variance. *Journal of the American Statistical Association* **32**: 675–701.

Gionfriddo, J.P. and Krausman, P.R. (1986). Summer habitat use by mountain sheep. *Journal of Wildlife Management* **50**: 331–6.

Giroux, J.F. and Bedard, J. (1988). Use of bulbrush marshes by greater snow geese during staging. *Journal of Wildlife Management* **52**: 415–20.

Green, E.L. (1973). Location analysis of prehistoric Maya sites in British Honduras. *American Antiquity* **38**: 279–93.

Grover, K.E. and Thompson, M.J. (1986). Factors influencing spring feeding site selection by elk in Elkhorn mountains, Montana. *Journal of Wildlife Management* **50**: 466–70.

Hamley, J.M. and Regier, H.A. (1973). Direct estimates of gillnet selectivity to walleye (*Stizostedion vitreum vitreum*). *Journal of the Fisheries Research Board of Canada* **30**: 817–30.

Harris, W.F. (1986). *The Breeding Ecology of the South Island Fernbird in Otago Wetlands*. Ph.D. Thesis, University of Otago, Dunedin, New Zealand.

Heisey, D.M. (1985). Analyzing selection experiments with log-linear models. *Ecology* **66**: 1744–8.

Hess, A.D. and Swartz, A. (1940). The forage ratio and its use in determining the food grade of streams. *Transactions of the North American Wildlife Conference* **5**: 162–4.

Hobbs, N.T. and Bowden, D.C. (1982). Confidence intervals on food preference indices. *Journal of Wildlife Management* **46**: 505–7.

Hobbs, N.T. and Hanley, T.A. (1990). Habitat evaluation: do use/availability data reflect carrying capacity? *Journal of Wildlife Management* **54**: 515–22.

Hohf, R.S., Ratti, J.T. and Croteau, R. (1987). Experimental analysis of winter food selection by spruce grouse. *Journal of Wildlife Management* **51**: 159–67.

Hohman, W.L. (1985). Feeding ecology of ringnecked ducks in northwestern Minnesota. *Journal of Wildlife Ecology* **49**: 546–57.

Holmes, R.T. and Robinson, S.K. (1981). Tree species preference of foraging insectivorous birds in a Northern hardwoods forest. *Oecologia* **48**: 31–5.

Hudgins, J.E., Storm, G.L. and Wakely, J.S. (1985). Local movements and diurnal–habitat selection by male American woodcock in Pennsylvania. *Journal of Wildlife Management* **49**: 614–9.

Huegel, C.N., Dahlgren, R.B. and Gladfelter, H.L. (1986). Bedsite selection by white–tailed deer fawns in Iowa. *Journal of Wildlife Management* **50**: 474–80.

Hurlbert, S.H. (1984). Pseudoreplication and the design of ecological field experiments. *Ecological Monographs* **54**: 187–211.

Iverson, G.C., Vohs, P.A. and Tacha, T.C. (1985). Habitat use by sandhill cranes wintering in western Texas. *Journal of Wildlife Management* **49**: 1074–83.

Ivlev, V.S. (1961). *Experimental Ecology of the Feeding of Fishes*. Yale University Press, New Haven.

Jacobs, J. (1974). Quantitative measurement of food selection: a modification of the forage ratio and Ivlev's electivity index. *Oecologia* **14**: 413–17.

Johnson, D.H. (1980). The comparison of usage and availability measurements for evaluating resource preference. *Ecology* **61**: 65–71.

Jolicoeur, P. and Brunel, P. (1966). Application du diagramme hexagonal a l'etude de la selection de ses proies par la Morue. *Vie Milieu B, Oceanography* **17(1-B)**: 419–33.

Kalmback, E.R. (1934). Field observations in economic ornithology. *Wilson Bulletin* **46**: 73–90.

Keating, K.A., Irby, L.R. and Kasworm, W.F. (1985). Mountain sheep winter food habits in the upper Yellowstone Valley. *Journal of Wildlife Management* **49**: 156–61.

Kincaid, W.B. and Bryant, E.H. (1983). A geometic method for evaluating the null hypothesis of random habitat utilization. *Ecology* **64**: 1463–70.

Kohler, C.C. and Ney, J.J. (1982). A comparison of methods for quantitative analysis of feeding selection of fishes. *Environmental Biology of Fishes* **7**: 363–8.

Krueger, W.C. (1972). Evaluating animal forage preference. *Journal of Range Management* **25**: 471–5.

Lachenbruch, P.A. (1976). Analysis of data with clumping at zero. *Biometrical Journal* **18**: 351–6.

Lagory, M.K., Lagory, K.E. and Taylor, D.H. (1985). Winter browse availability and use by white-tailed deer in southeastern Indiana. *Journal of Wildlife Management* **49**: 120–4.

Larsen, D.L. and Bock, C.E. (1986). Determining avian habitat preference by bird-centered vegetation sampling. In *Wildlife 2000: Modelling Habitat Relationships of Terrestrial vertebrates* (eds J. Verner, M.L. Morrison and C.J. Ralph), pp. 37–43. University of Wisconsin Press, Madison.

Laymon, S.A., Salwasser, H. and Barrett, R.H. (1985). *Habitat Suitability Index Models: Spotted Owl*. Biological Report 82(10.113), United States Fish and Wildlife Service, Biological Services Program, Washington, D.C.

Lechowicz, M.J. (1982). The sampling characteristics of electivity indices. *Oecologia* **52**: 22–30.

Loehle, C. and Rittenhouse, L.R. (1982). An analysis of forage preference indices. *Journal of Range Management* **35**: 316–19.

Manly, B.F.J. (1973). A linear model for frequency–dependent selection by predators. *Researches on Population Ecology* **14**: 137–50.

Manly, B.F.J. (1974). A model for certain types of selection experiments. *Biometrics* **30**: 281–94.

Manly, B.F.J. (1985). *The Statistics of Natural Selection on Animal Populations*. Chapman and Hall, London.

Manly, B.F.J. (1986). *Multivariate Statistical Methods: a Primer*. Chapman and Hall, London.

Manly, B.F.J. (1990). *Stage-Structured Populations: Sampling, Analysis and Simulation*. Chapman and Hall, London.

Manly, B.F.J. (1991). *Randomization and Monte Carlo Methods in Biology*. Chapman and Hall, London.

Manly, B.F.J., Miller, P. and Cook, L.M. (1972). Analysis of a selective predation experiment. *American Naturalist* **106**: 719–36.

Marcum, C.L. (1975). *Summer–Fall Habitat Selection and Use by a Western Montana Elk Herd*. Ph.D. Thesis. University of Montana, Missoula.

Marcum, C.L. and Loftsgaarden, D.O. (1980). A non–mapping technique for studying habitat preferences. *Journal of Wildlife Management* **44**: 963–8.

McCorquodale, S.M., Raedeke, K.J. and Taber, R.D. (1986). Elk habitat use patterns in the shrub-steppe of Washington. *Journal of Wildlife Management* **50**: 664–9.

McCullagh, P. and Nelder, J.A. (1989). *Generalized Linear Models, 2nd edn.* Chapman and Hall, London.

McIntosh, A.R., Townsend, C.R. and Crowl, T.A. (1992). Competition for space between introduced brown trout (*Salmo trutta* L.) and a native galaxiid (*Galaxias vulgaris* Stokell) in a New Zealand stream. *Journal of Fish Biology* **41**: 63–81.

Mielke, P.W. (1986). Non-metric statistical analyses: some metric alternatives. *Journal of Statistical Planning and Inference* **13**: 377–87.

Millar, R.B. and Walsh, S.J. (1992). Analysis of Trawl Selectivity studies with an application to Frouser trawls. *Fisheries Research* **13**: 205–20.

Morrison, D.F. (1976). *Multivariate Statistical Methods, 2nd edn.* McGraw-Hill, New York.

Munro, H.L. and Rounds, R.C. (1985). Selection of artificial nest sites by five sympatric passerines. *Journal of Wildlife Management* **49**: 264–78.

Murphy, R.K., Payne, N.F. and Anderson, R.K. (1985). White–tailed deer use of an irrigated agricultural–grassland complex in central Wisconsin. *Journal of Wildlife Management* **49**: 125–8.

Nams, V.O. (1989). Effects of radiotelemetry error on sample size and bias when testing for habitat selection. *Canadian Journal of Zoology* **67**: 1631–6.

Nelson, J.R. (1978). Maximizing mixed animal species stocking rates under proper–use management. *Journal of Wildlife Management* **42**: 172–4.

Nelson, M.E. (1979). Home range location of white-tailed deer. United States Forest Service Research Paper NC–173.

Neu, C.W., Byers, C.R. and Peek, J.M. (1974). A technique for analysis of utilization-availability data. *Journal of Wildlife Management* **38**: 541–5.

Nudds, T. (1980). Forage 'preference': theoretical considerations of diet selection by deer. *Journal of Wildlife Management* **44**: 735–9.

Nudds, T. (1982). Theoretical considerations of diet selection by deer: a reply. *Journal of Wildlife Management* **46**: 257–8.

Owen–Smith, N. and Cooper, S.M. (1987). Assessing food preferences of ungulates by acceptability indices. *Journal of Wildlife Management* **51**: 372–78.

Paloheimo, J.E. (1979). Indices of food type preference by a predator. *Journal of the Fisheries Research Board of Canada* **36**: 470–3.

Palomares, F. and Delibes, M. (1992). Data analysis and potential bias in radio-tracking studies of animal use. *Acta Oecologia* **13**: 221–6.

Parsons, B.G.M. and Hubert, W.A. (1988). Influence of habitat availability on spawning site selection by kokanees in streams. *North American Journal of Fisheries Management* **8**: 426–31.

Pearre, S. (1982). Estimating prey preference by predators: uses of various indices, and a proposal of another based on χ^2. *Canadian Journal of Fisheries and Aquatic Sciences* **39**: 914–23.

Peek, J.M. (1986). *A Review of Wildlife Management*. Prentice-Hall, New Jersey.

Petersen, M.R. (1990). Nest–site selection by emperor geese and cackling Canada geese. *Wilson Bulletin* **102**: 413–26.

Petrides, G.A. (1975). Principal foods versus preferred foods and their relations to stocking rate and range condition. *Biological Conservation* **7**: 161–9.

Pietz, P.J. and Tester, J.R. (1982). Habitat selection by sympatric spruce and ruffed grouse in north central Minnesota. *Journal of Wildlife Management* **46**: 391–403.

Pietz, P.J. and Tester, J.R. (1983). Habitat selection by snowshoe hares in north central Minnesota. *Journal of Wildlife Management* **47**: 686–96.

Pinkas, L., Oliphant, M.S. and Iverson, I.L.K. (1971). *Food Habits of Albacore, Bluefin Tuna, and Bonito in California Waters*. Fisheries Bulletin **152**, California Department of Fisheries and Game.

Popham, E.J. (1944). A study of the changes in an aquatic insect population using minnows as the predators. *Proceedings of the Zoological Society of London* **A114**: 74–81.

Popham, E.J. (1966). An ecological study of the predatory action of the three spine stickleback (*Gasterosteus aculeatus* L.). *Archives for Hydrobiology* **62**: 70–81.

Porter, W.F. and Church, K.E. (1987). Effects of environmental pattern on habitat preference analysis. *Journal of Wildlife Management* **51**: 681–5.

Porter, M.L. and Labisky, R.F. (1986). Home range and foraging habitat of red-cockaded woodpeckers in northern Florida. *Journal of Wildlife Management* **50**: 239–47.

Prevett, J.P., Marshall, I.F. and Thomas, V.G. (1985). Spring foods of snow and Canada geese at James Bay. *Journal of Wildlife Management* **49**: 558–63.

Pyke, G.H., Pulliam, H.R. and Charnov, E.L. (1977). Optimal foraging: a selective review of theory and tests. *Quarterly Review of Biology* **52**: 137–54.

Quade, D. (1979). Using weighted rankings in the analysis of complete blocks with additive block effects. *Journal of the American Statistical Association* **74**: 680–3.

Rachlin, J.W., Pappantoniou, A. and Warkentine, B.E. (1987). A bias estimator of the environmental resource base in diet preference studies with fish. *Journal of Freshwater Ecology* **4**: 23–31.

Rapport, D.J. (1980). Optimal foraging for complementary resources. *American Naturalist* **116**: 324–46.

Rapport, D.J. and Turner, J.E. (1970). Determination of predator food preferences. *Journal of Theoretical Biology* **26**: 365–72.

Raley, C.M. and Anderson, S.H. (1990). Availability and use of arthropod food resources by Wilson's warblers and Lincoln's sparrows in southeastern Wyoming. *Condor* **92**: 141–50.

Ready, R.C., Mills, E.L. and Confer, J.L. (1985). A new estimator of, and factors influencing, the sampling variance of the linear index of food selection. *Transactions of the American Fisheries Society* **114**: 258–66.

Reed, T.E. (1969). Genetic experiences with a general maximum likelihood estimation program. In *Computer Applications in Genetics* (ed N.E. Morton), pp. 27–9. University of Hawaii Press, Honolulu.

Reed, T.E. and Schull, W.J. (1968). A general maximum likelihood estimation program. *American Journal of Human Genetics* **20**: 579–80.

Rexstad, E.A., Miller, D.D., Flather, C.H., Anderson, E.M., Hupp, J.W. and Anderson, D.R. (1988). Questionable multivariate statistical inference in wildlife habitat and community studies. *Journal of Wildlife Studies* **52**: 794–8.

Rich, T. (1986). Habitat and nest–site selection by borrowing owls in the sagebrush steppe of Idaho. *Journal of Wildlife Management* **50**: 548–55.

Rolley, R.E. and Warde, W.D. (1985). Bobcat habitat use in southeastern Oklahoma. *Journal of Wildlife Management* **49**: 913–20.

Rondorff, D.W., Gray, G.A. and Fairley, R.B. (1990). Feeding ecology of subyearling Chinook Salmon in riverine and reservoir habitats of the Columbia River. *Transactions of the American Fisheries Society* **119**: 16–24.

Rosenzweig, M.L. (1981). A theory of habitat selection. *Ecology* **62**: 327–35.

Roy, L.D. and Dorrance, M.J. (1985). Coyote movements, habitat use, and vulnerability in central Alberta. *Journal of Wildlife Management* **49**: 307–13.

Ryder, T.J. (1983). *Winter Habitat Selection by Pronghorn in South Central Wyoming*. M.S. Thesis, Department of Zoology and Physiology, University of Wyoming.

Savage, R.E. (1931). The relation between the feeding of the herring off the east coast of England and the plankton of the surrounding waters. Fishery Investigation, Ministry of Agriculture, Food and Fisheries, Series 2, **12**: 1–88.

Schoen, J.W. and Kirchhoff, M.D. (1985). Seasonal distribution and home range patterns of Sitka black-tailed deer on Admiralty Island, southeast Alaska. *Journal of Wildlife Management* **49**: 96–103.

Scott, A. (1920). Food of Port Erin mackerel in 1919. *Proceedings and Transactions of the Liverpool Biological Society* **34**: 107–11.

Seber, G.A.F. (1984). *Multivariate Observations*. Wiley, New York.

Sheppard, P.M. (1951). Fluctuations in the selective value of certain phenotypes in the polymorphic land snail *Cepaea nemoralis* (L.). *Heredity* **5**: 125–34.

Smith, R.E., Hupp, J.W. and Ratti, J.T. (1982). Habitat use and home range of grey partridge in eastern south Dakota. *Journal of Wildlife Management* **46**: 580–7.

Stauffer, D.F. and Peterson, S.R. (1985). Ruffed and blue grouse habitat use in southeastern Idaho. *Journal of Wildlife Management* **49**: 459–66.

Stinnett, D.P. and Klebenow, D.A. (1986). Habitat use of irrigated lands by California quail in Nevada. *Journal of Wildlife Management* **50**: 368–72.

Strauss, R.E. (1979). Reliability estimates for Ivlev's electivity index, the forage ratio, and a proposed linear index of food selection. *Transactions of the American Fisheries Society* **108**: 344–52.

Talent, L.G., Krapu, G.L. and Jarvis, R.L. (1982). Habitat use by mallard broods in south central North Dakota. *Journal of Wildlife Management* **46**: 629–35.

Thomas, D.L., and Taylor, E.J. (1990). Study designs and tests for comparing resource use and availability. *Journal of Wildlife Management* **54**: 322–30.

Thomasma, L.E., Drummer, T.D. and Peterson, R.O. (1991). Testing the Habitat Suitability Index for the fisher. *Wildlife Society Bulletin* **19**: 291–7.

Thompson, J.N. (1988). Evolutionary ecology of the relationship between oviposition preference and performance of offspring in phytophagus insects. *Entomologia Experimentalis et Applicata* **47**: 3–14.

United States Fish and Wildlife Service (1981). *Standards for the Development of Suitability Index Models*. Ecology Service Manual 103, United States Fish and Wildlife Service, Division of Ecological Services, Washington, D.C.

van Horne, B. (1983). Density as a misleading indicator of habitat quality. *Journal of Wildlife Management* **47**: 893–901.

Vanderploeg, H.A. and Scavia D. (1979a). Two electivity indices for feeding with special reference to zooplankton grazing. *Journal of the Fisheries Research Board of Canada* **36**: 362–5.

Vanderploeg, H.A. and Scavia, D. (1979b). Calculation and use of selectivity coefficients of feeding: zooplankton grazing. *Ecological Modelling* **7**: 135–49.

Walsh, S.J., Millar, R.B., Cooper, C.G. and Hickey, W.M. (1992). Codend selection in American plaice: diamond versus square mesh. *Fisheries Research* **13**: 235–54.

Werner, E.E. and Hall, D.J. (1974). Optimal foraging and the size selection of prey by the bluegill sunfish (*Lepomis macrochirus*). *Ecology* **55**: 1042–52.

White, G.C. and Garrott, R.A. (1990). *Analysis of Wildlife Radio–Tracking Data*. Academic Press Inc., New York.

White, R.G. and Trudell, J. (1980). Habitat preference and forage consumption by reindeer and caribou near Atkasook, Alaska. *Arctic and Alpine Research* **12**: 511–29.

Whitham, T.G. (1980). The theory of habitat selection: examined and extended using Pemphigus aphids. *American Naturalist* **115**: 449–66.

Wiens, J. A. (1981). Scale problems in avian censusing. In *Estimating Numbers of Terrestrial Birds* (eds C.J. Ralph and J.M. Scott), pp. 513–21. Studies in Avian Biology 6, Cooper Ornithological Society.

Wong, B. and Ward, F.J. (1972). Size selection of *Daphnia publicaria* by yellow perch (*Perca flavescens*) fry in West Blue Lake, Manitoba. *Journal of the Fisheries Research Board of Canada* **29**: 1761–4.

Zaret, T.M. and Kerfoot, W.C. (1975). Fish predation on *Bosmina longirostris:* body-size selection versus visibility selection. *Ecology* **56**: 223–37.

Author Index

Subject Index